Environmental Deterioration in the Soviet Union and Eastern Europe

edited by
Ivan Volgyes

Environmental Deterioration in the Soviet Union and Eastern Europe

Praeger Special Studies in International Politics and Government

Praeger Publishers   New York   Washington   London

Library of Congress Cataloging in Publication Data

Volgyes, Ivan, 1936-
    Environmental deterioration in the Soviet Union and
Eastern Europe.

    (Praeger special studies in international politics
and government)
    1. Pollution—Russia.   2. Pollution—Europe,
Eastern.   I. Title.
TD186.5.R9V6   363.6   74-3141
ISBN 0-275-08920-7

PRAEGER PUBLISHERS
111 Fourth Avenue, New York, N.Y. 10003, U.S.A.
5, Cromwell Place, London SW7 2JL, England

Published in the United States of America in 1974
by Praeger Publishers, Inc.

ACKNOWLEDGMENTS

Most of the papers included in this volume were presented first in summary form at the First International Congress of the Society for Engineering Science held in Tel Aviv, Israel, June 12-17, 1972. Abstracts of the papers were included in the Proceedings of the Congress and an earlier version of the introductory framework article by the editor was published in the volume that grew out of the convention: E. S. Barrekette, ed., Pollution: Engineering and Scientific Solutions (New York and London: Plenum Press, 1973). The editor gratefully acknowledges the thoughtful invitation of Professor Ervin Y. Rodin of Washington University, St. Louis, Missouri, to include on the program a session dealing with political considerations, and the assistance given by Professor W. A. Douglas Jackson of the University of Washington, Seattle, Washington, to make this session a success.

The editor would also like to express his gratitude to Mrs. E. S. Dunklau who typed and retyped most of the papers and to Mrs. Mary Volgyes, who gave invaluable assistance in the editorial work for this volume.

CONTENTS

LIST OF MAPS

LIST OF TABLES AND FIGURES

Environmental Deterioration in the Soviet Union and Eastern Europe

1

POLITICS AND POLLUTION
IN WESTERN AND
COMMUNIST SOCIETIES
Ivan Volgyes

The waters of Lake Baikal are nearly as polluted as those of
the Great Lakes. Smoke chokes people in Budapest and Tokyo, Mexico
City and New York, and automobile pollution has destroyed the clean
air of small provincial towns in many countries. These are the symp-
toms of the industrial age. Industrialization is destroying humanity
regardless of race or sex, ideology or religion. Industrialization,
modernization, and urbanization have begun to exhibit serious tendencies of failure just when the lofty dream of bringing health and
abundance to the people of the world seems within reach.

The ills of society, of course, are not new. Pollution has been
with man at least since he began to live like a civilized being. The
Jerusalem of Biblical days probably was a virtual dungheap by modern
standards. The ports of ancient Greece were so polluted that one
had to climb over the refuse to see the ships come and go. The streets
of Rome were rat-infested. Smoke from cooling fires darkened the
villas of Pompeii. The grime, soot, and dirt of Paris were worse in
the sixteenth century than in recent times, and the great fire in London
destroyed a decayed inner city far more filthy and polluted than it
was before the air pollution control efforts began in the 1960s. The
center of New York in the 1880s was almost as unhealthy to live in
as it is today.

Until the second half of the twentieth century, man had never
really begun to clean up the environment. Although the list of people

An earlier version of this article, first presented as an invited
paper at the International Conference of the Society for Engineering
Science in Tel Aviv, Israel, June 12-17, 1972, is reproduced in E. S.
Barrekette, ed., Pollution: Engineering and Scientific Solutions (New
York and London: Plenum Press, 1973), pp, 680-88.

interested in the topic included Dante and Machiavelli, Erasmus and
Kepler, Bentham and Marx, little was ever accomplished. All of the
efforts to deal with pollution failed, because those pursuing them had
limited ways of exerting political pressure on the rulers of the time.
Even today, politics is the key to implementing technical solutions to
the problems posed by a constantly deteriorating environment, and
regardless of the system of government, it is primarily political considerations that define the possible limits on the control of our physical
environment.

The problem of controlling environmental deterioration first
became an issue in the developed capitalist democracies of the West
with their technological societies and advanced capacities for wreaking
havoc on the natural surroundings; but the very philosophies that,
since the onset of the industrial revolution, have guided the economic
growth of these societies make the problem of combating pollution
extremely difficult. What appears to be needed is strict government
control of an area in which governments traditionally have wished to
exert a minimum of interference. Thus, the political means of solving
the problems of the deteriorating environment in these countries is
essentially placed in the realm of interest-group politics. In each of
these states the role of the government in environmental control, to
a lesser or greater extent, is that of an arbiter among interest groups.
The demand articulation of these interest groups ranges from a very
significant level capable of effecting changes to minimal, poorly
financed, badly coordinated efforts. Thus, for example, in the United
States the professional lobbies working for such giants as United
States Steel or Dupont are more likely to be able to muster political
strength than the disparate local groups fighting for environmental
control measures. In addition, the legislators in the democratic
countries frequently have reelection as a primary concern. They are
likely to pay more attention to the local economic importance of a
factory than to the apparently less immediate needs of the citizenry
for clean water or breathable air.

Legislators, however, are slowly being compelled by increasingly
more powerful citizens' lobbies to pay attention to the environment
that industrialization has destroyed. The high mercury content of
polluted lakes threatens the children of legislators and of industrialists
and bankers, as well as those of the workers and middle-class families.
It thus becomes their common concern to control the pollutants, and
even if their efforts result in a temporary loss in profits for the com-
panies, public pressure has begun to make it mandatory that such
controls be established.

In the advanced Western societies the problem is complicated
by the fact that the economic system is based on the notion of profit.
The costs of environmental measures thus may be tagged onto the

2

price of products, causing prices to advance steadily. The argument advanced by the industries is: "If you want cleaner air, water, or nonpolluting cars, you, the customer, must pay for it." Implicit in this argument, unfortunately, is the assumption that there is a certain "natural right" of the producer to pollute the environment in exchange for producing his wares.

In the United States the political solutions to this problem will require a greater role for the federal government as the representative of all of the people and of the national interest. It now is expected to supply the needed impetus for change. The government traditionally has adopted two methods of combating the problem. On the one hand it can supply subsidies to factories, thereby combating the inflationary trend, and on the other hand it can enforce federal standards of maximum allowable pollution. The first approach saddles the nation with higher taxes, thereby forcing all the people of the United States to pay for profits that would accrue to only a small percentage of the citizenry. This approach puts the burden of paying for the cleaning up of the environment back on the populace. The second approach, enforcing strict federal standards, also poses difficulties. First, federal standards, as written by Congress or legislatures, always are results of compromise. They are never optimum solutions, and at best they require only those changes that would assure minimally acceptable conditions. Congress can never view the problem of ecological deterioration without attempting to reap clear political profit for its members. Second, even when established, federal standards have been difficult to enforce. Industrial pollution in West Virginia mines, for example, has not been carefully controlled in spite of the fact that federal legislation has been enacted since 1969. Large steel plants, chemical firms, and paper milling industries find loopholes in such legislation with relative ease, and enforcement would require so many federal agents at various jobs that even the burgeoning civil service in the United States would not be able to undertake such a task.

Another significant problem with pollution control in the United States relates to differences in the public view of national priorities. While the prosperous middle class views the problem of pollution as an important issue, this view is not shared by the residents of the inner cities and ghettoes. To the residents of these areas, the problem of pollution is not a primary concern. The question of unbreathable air comes secondary to the building of new hospitals. The problem of industries belching smoke is not as important as having factories that will provide jobs for the unemployed and pay decent wages. To the ghetto dwellers cleaning up Lake Erie is a minor concern in comparison to the need for replacing the rat-infested, urban dwellings of Chicago.

Other industrial democracies have environmental problems similar to those of the United States. All of them are groping for a solution to automobile pollution, industrial waste, and water pollution, but in many of these countries the maintenance of a modern industrialized economy is a primary concern; the need for environmental controls has had to be secondary. The tragic point, however, is that many of the people themselves are not aware of being poisoned daily. Some simply accept the fact that the price of industrialization is pollution and are meekly looking toward their governments to minimize the dangers while maximizing their standards of living.

In all of these industrialized democracies some control measures are being implemented and a greater and greater consciousness is emerging among the people to demand that the government enact more effective controls. Government is a more powerful instrument vis-à-vis industry in many of these states than in the United States. In addition, the national concerns usually are less diffuse, the urban areas have not been allowed to deteriorate into slums, and therefore the problems do not seem to loom as large as they do in the United States. International agencies, too, are attacking the problem in Western Europe. Within the Common Market, for example, a new framework for pollution control exists, and perhaps within this framework some significant action can be undertaken in the 1970s to reverse the environmental deterioration of Western Europe.

The problem of pollution and the implementation of control measures hinge on political solutions in all of the industrialized democracies. The question of the fine balance between the powerful industries and the governments, or between the economic and ecological interests of the states concerned, can only be reached through political means. If any effective action is to be taken by these states, governments and opposition parties alike must begin to incorporate in their programs a recognition of the problem of pollution and means for reversing deterioration of the environment. In this regard, popular pressure will be mandatory in combating the influence of the powerful industrial interest groups.

It would appear that in the Communist states the specific problems of the Western democracies would not pose obstacles to the control of environmental deterioration. These societies industrialized decades, or even centuries, later than their Western counterparts, and it would seem that some of the Western experiences could serve as warnings to them as they structure their new societies. Eastern Europe, particularly, could have benefited from the Western examples since industrialization in parts of this area began as late as the end of World War II. In addition, the traditional reluctance of Western governments to interfere too drastically in the economic sphere seems to pose no problem in the Communist states. Here we are dealing with a group of states whose governments exercise a strong control over the polity and where industry, if not state-owned, is at least state-controlled. Supposedly, thus, we are concerned with states where the government is relatively free to deal with the industries as it pleases and therefore is able to insist on the quality and character of the environment. This generalization ought to apply clearly to the USSR and to the other Communist countries as well as to a number of states with strong authoritarian regimes. Yet there exists such a complex situation in these states that such generalizations are not warranted. Particularly in the Communist countries, the governments have come to power with a mandate to industrialize, to lead the people to the promised land of a better standard of living. In other words, the ideologies have promised the material abundance that exists in capitalist countries. The image of the good life in capitalist states remains extremely powerful: progress means industrialization; it demands more and more factories and hydroelectric plants, and eventually it means having private automobiles.

Thus, the governments have placed tremendous national efforts on industrialization, and the results have been pollution of the air and destruction of the rivers and lakes. Modernization became the key to the success of these regimes and environmental concerns have had to be placed near the bottom of the political concerns. A sardonic Hungarian bitterly remarked, "We have already reached the Western capitalist level in the amount of pollution we produce." Lake Baikal is nearly ruined from pollution, the Danube is choked with millions of tons of industrial waste, oxygen-dispensing stations have been set up in Budapest during the winter, and the factories of Galati and Nowa Huta and the other industrial centers of the Communist states belch smoke with little evidence of government or popular concern. Coal heating is slowly being replaced with gas heat in some places, but gas and oil cost more money—a critical factor in poor countries lacking natural resources. Several of the states have written standards for control of industrial pollution and for relocation of industries from the centers of urban areas, but the essential poverty of the countries drastically limits the options and the possible solutions. Lacking money, resources, and technology, their hands are tied no matter how strong the national commitment. In addition, the demand for more industries never ceases. Defense requirements call for ever-growing amounts of iron and steel, while the scarce quantities of consumer goods must be maintained to keep the population happy.

Obviously, the political control of production and pollution are vastly different from the practices of the industrialized Western democracies. In these industrializing systems it is the state, the government, that pollutes and is reluctant to limit its productive capacities. Specifically, where single-party systems rule, all power

5

to control the environment is theoretically vested in the regime's hands. The rulers are free to decide whether to pollute or not to pollute. That freedom, of course, is vastly encumbered by the demands of the population to keep up with the luxuries of the developed countries Few are willing to accept decreased availability of consumer goods and a lessened tempo of industrialization, even if it would mean cleaner air.

During the late 1960s, however, a few voices of dissent were beginning to be heard within some of the ruling hierarchies in these systems. The pollution of Lake Baikal has become a national scandal in the USSR and some of the younger economists and city planners have begun to question the utility of headlong industrialization. Primitive ecological interest-group articulation has begun to be noticeable in Hungary, Yugoslavia, and even within the Soviet economic elite. But so far in all of these states a national commitment to severely limit the pollution is missing, and in spite of the governments' theoretical abilities to carry out wholesale reforms, a commitment to such ecological reform is nonexistent.

It must be recognized, however, that even if the authoritarian regimes of the industrializing Communist world decide to make the necessary commitment to limit environmental deterioration, they are likely to face problems similar to or more difficult than those faced by the West, for in these countries the advanced technology and relative richness of the highly industrialized societies is missing, and the concern for continued development places limitations on the money available to combat environmental deterioration. Moreover, the relative inexperience of the Communist states with interest-group politics and the demands to assign political values and priorities to each course of action advocated by widely divergent groups tend to slow down control measures. Thus, although monocratic regimes theoretically are more able to influence the control of a deteriorating environment, the practical limitations on government enforcement are going to be considerable.

The obvious philosophical implications of the limitations faced by the two major forms of government in the modern world are depressing, because they suggest that mankind has sentenced itself to live in ever-increasing industrial waste. Yet, soon the question will move out of the realm of philosophy and into the simple area of man's need to survive, because while mankind, particularly the urban dweller, has been faced with environmental deterioration since the dawn of history, modern technology has intensified the problems a millionfold. First, we are no longer dealing with just filth and human refuse. Now we have plastics and detergents that are not biodegradable—that once produced are destined to be permanent additions to the ecological system. We also are dealing with the chemical poisoning of food

supplies with hidden killers—additives that may affect the unborn
generations with mutations and deformation. The affect of sunlight
on chemical additives in the air is another modern nightmare with
unkown, and perhaps terrifying, implications. These problems, coupled
with the sheer increase in our capabilities to produce smoke, chemical
wastes, water pollutants, and garbage are overwhelming. Unless inno-
vative political means for dealing with the problem are devised soon,
we may destroy not only the present political systems, but also civilization and perhaps mankind itself.

2

THE SOVIET RESPONSE
TO ENVIRONMENTAL DISRUPTION
Keith Bush

As in the developed nations of the West, the past decade has witnessed a growing awareness in the USSR of the scale and intensity of environmental disruption, together with recognition by the Soviet authorities of the need for more meaningful measures to curb and to reduce the harmful ecological consequences of modernization. During the past few years, impressive and comprehensive legislation to counter environmental disruption has been enacted. The next decade will show how thoroughly this will be implemented and what further measures must be adopted.

Since world attention was focused to an unprecedented extent upon the environmental hazards stemming from industrialization, motorization, land improvement, and other facets of modernization at the United Nations Conference on the Human Environment, held in June 1972 in Stockholm, it is perhaps timely to examine the Soviet response to this threat that recognizes no national frontiers. This short review will not seek to prove the existence of air and water pollution or soil erosion in the USSR, for this is no longer disputed by serious observers. Instead it will attempt to distinguish some of the factors responsible for a serious degree of environmental disruption in the USSR and the prophylactic measures undertaken there that differ, wholly or in part, from those in other countries. The achievements and the shortcomings of the Soviet experience may be instructive for concerned observers from both mixed "capitalist" economies as well as from those less-developed countries that still have to select a development model.

This study has benefited from helpful comments by Mrs. Siri Kalnins, Professor Norton Dodge, and Professor Marshall Goldman.

THE SIZE OF THE PROBLEM

The incidence and the scale of environmental disruption in the USSR have been diligently and competently chronicled in many Western studies, notably in those of Marshall Goldman,[1] and thus only some salient examples will be cited here.

Insofar as the extent and overall intensity of pollution can be quantified, most observers would agree that the natural resources of the USSR have not been affected to the same degree as in, say, the United States, Japan, or Western Europe for the simple reason that most sectors of the Soviet economy are at a lower stage of development and because the population density is less. For instance, the primary source of air pollution in the United States is the automobile: in 1965 it deposited an estimated 86 million tons of pollutants in the atmosphere, representing some 60 percent of total air pollutants.[2] With a current Soviet stock of 1.5 million privately owned automobiles,[3] against a U.S. inventory approaching 85 million, the USSR's lag in this respect is clearly an environmental plus, even if its fewer cars use less refined fuel and emit a greater share of this into the air. The population density is roughly one-half of that of the United States, although the bulk of the Soviet population is concentrated west of the Urals.

If the global extent of environmental disruption in the USSR is at present less than in most Western developed economies the expected growth of "ecological demand," i.e., man's demands on the environment such as the extraction of resources and the return of waste, will be several times higher than overall economic growth (projected for the 1970s by Western analysts at about 4 percent per annum)[4] that underlines the urgency for effective countermeasures. Moreover, as in the West, environmental disruption tends to be highly localized in the USSR. In terms of deleterious effects upon human and animal health and well-being, water pollution presents the gravest threat, followed by air pollution. In terms of economic damage, the erosion of soil by wind and water probably accounts for the greatest losses.

Water Management and Water Pollution

The territory of the USSR comprises some 16 percent of the world's land mass, but includes only 11 percent of its fresh water supplies. The distribution of this share is unfavorable, with only 12 percent of the total available to the densely populated regions of European Russia, the Caucasus, and Central Asia. By the end of this century, the urban population will have grown to an estimated 216 million, while the per capita consumption of fresh water for urban dwellers at that time is being planned at about 500 liters per day, that is, a 100 to 150 percent increase on present levels.[5] Thus, fresh water is already relatively scarce and will become scarcer. Yet the supply of this vital commodity is being reduced and endangered not only by industrial and sewage pollution, but also by the diversion and depletion of existing bodies of water by vast land improvement schemes

Inspired primarily by the harvest failures of 1963 and 1965 and by the subsequent humiliation of massive imports of Western food grains, successive Soviet leaders have announced grandiose projects for reclaiming and irrigating tens of millions of hectares of swampy and arid lands. The present leadership's plans were unveiled at the March 1965 plenum of the Central Committee of the Communist party of the Soviet Union, elaborated upon at the May 1966 plenum, and reaffirmed at the October 1968 and July 1970 plenums. The current goal for irrigation is 21 million hectares by 1985.[6] Such plans envisage, inter alia, the diverting of the courses of the great Siberian rivers to the dry regions of Kazakhstan and Central Asia, half of the annual flow of the Amu-Daria and Syr-Daria rivers to be drawn off, and up to one-quarter of the Volga's water to be diverted to irrigate the Povolzhe. The ecological and climatic consequences of such projects cannot be predicted with accuracy. A part of the Arctic ice cap might melt, the climate of the Northern Hemisphere might be altered, while the irrigation of millions of hectares could increase evaporation to such an extent that existing rainfall patterns might be changed. Yet despite protests at these colossal gambles with nature's water balance and demands for fuller prior scientific investigation from Soviet and Western specialists,[7] work on diverting Siberian river courses was under way as this chapter was being written. From the junction of the Tobol and Irtysh, water is to be pumped to Kazakhstan, to a reservoir in the Aral lowlands, and then to desert regions in the Kazakh, Uzbek, and Turkmen SSRs.[8]

Apart from the diversion of water for agricultural purposes, the degradation of the existing bodies of water and flows from untreated or inadequately treated industrial effluents and sewage have been extensively documented in the Soviet press. Perhaps the most dramatic and tragic manifestation of this has been the drop in the sturgeon catch from the Caspian Sea, from over 50,000 tons a year in the seventeenth century to less than 10,000 tons a year now. Some 65,000 tons of oil—twice as much as the load of the Torrey Canyon—seep into Baku Bay each year.[9] Of course, the output not only of Soviet but also of Iranian beluga caviar has been adversely affected.

Without adequate facilities for recycling and reuse, the amount of water needed to dilute industrial effluents is already immense and is growing rapidly. Although they have not reached the level of pollution of, say, the Rhine or the Mississippi, the great rivers of the Soviet Union are being gradually turned into sewers: for instance, it has been estimated that the Volga alone carries half of the country's industrial effluents into the Caspian Sea.[10] The quantity of water required for Siberian industry in 1968 was set at nearly 150 million cubic meters per day. As industrial output in Siberia is planned to grow by 500 percent during the next 10 to 15 years, an amount equal to all the water that runs into the ocean from the Enisei and the Ob will be needed, unless radically new processes are introduced.[11]

Sewage facilities are often still primitive. A volume on the order of 25 cubic kilometers of untreated water is introduced each year into open reservoirs.[12] Until the mid-1960s, 300,000 to 400,000 cubic meters of raw sewage was flushed each day into the Moscow River alone.[13] This has reportedly been stopped, and "virtually all" domestic sewage is now purified, but the discharge of untreated industrial waste water into the river is expected to continue until 1976.[14]

Some Western environmentalists equate the flush toilet with the internal combustion engine as ecological enemy Number One: without it, they claim, the fresh water requirements of the average urban dweller could be cut to closer to 10 gallons rather than 100 gallons per day. The alternative, it might be countered, is mindbending. Yet the joys of a flush toilet are still not universal in the USSR: at the latest count (in 1968), 32 percent of state-owned urban housing was not equipped with sewage facilities and 29 percent was without running water.[15] The respective shares of all urban housing must have been greater. But all new apartments are being fully equipped in these respects and, as we have seen, the Soviet planners are counting upon the flush toilet to be with us still in the year 2000.

A spectacularly visible source of water pollution in many Western nations is the chemical detergent, and legislation has been passed, or is being prepared, in several countries stipulating that household and industrial detergents should be largely biodegradable, that is, capable of breaking down into smaller chemical molecules as a result of bacterial action. This particular consumer convenience has, however, only recently reached the Soviet market and few if any cases have been reported of water foaming from the tap or of streams with fine heads on them.

An increasingly common hazard, shared by the Soviet Union with other agricultural countries, is the pollution of lakes and rivers by agricultural chemicals. This can occur through drainage, but one of the worst cases of pollution occurred when dust storms blew fertilizer mixed with topsoil off of land in the Krasnodar Krai into the Sea of Azov.

As in the West, specially equipped vessels are used to remove floating oil slicks and refuse from harbor areas and coastal regions.[16]

In addition, Soviet biologists are breeding microbes that can "eat" oil: a recent report mentioned 37 species of zooplancton organisms that thrive on a diet of petroleum.[17]

Air Pollution

The USSR enjoys a clear advantage over most Western nations insofar as the quality of air in its cities is concerned. As has been mentioned, the principal pollutant of urban air in the West, the private automobile, is present only in relatively small numbers in the Soviet Union, with a current stock of 1.5 million units, that is, the level reached by the United States in mid-1913. This decade will witness the belated birth of the mass motoring age in the USSR, with planned sales of 2.6 million during the period 1971-75.[18] Here, then, is one field where the Russians can profit from the experience of the West and where, to judge from the technical literature, they are at least aware of the problems involved. Whether the planners are prepared to delay or to reduce production on environmental grounds, or to invest adequately in emission control devices, remains to be seen. Production models have not been affected so far.

At present, the relatively small number of automobiles do emit some rather noxious fumes into the atmosphere. This is mainly due to impure gasoline, to the existence of many very old and tired machines and to problems of servicing. However, with the debut of the Zhiguli, the output of higher-octane fuel has been stepped up, while servicing facilities can only improve. Moreover, the current drive to expand export markets for Soviet-made automobiles will presumably oblige manufacturers to conform to existing and projected exhaust emission standards.

Research has been in progress for several years on emission control devices. One type of catalytic neutralizer has been tested on passenger cars and buses in Alma-Ata since 1967: with vehicles moving at speeds of up to 30 to 40 kilometers per hour, this reportedly neutralizes nine-tenths of the waste gases.[19] One major drawback of this device is its short effective life of 5,000 to 6,000 kilometers. Another stringent requirement is for a universally applicable catalyst that can withstand temperatures ranging from the -60° C encountered in Siberia to an operating temperature of 1,000° C. A further prototype which has been tested successfully on Volga automobiles, cuts in automatically when combustion is incomplete, that is, during braking or coasting.[20] Fuel injection for motor cars is in its infancy, with the first production models scheduled for 1973.[21]

As in the West, attempts have been made at many centers and for many years to develop a viable battery-driven vehicle. The results have been much the same: the models designed so far have an in-adequate radius of operation, poor acceleration, tend to weigh too much, and are extremely expensive.[22] One interesting compromise solution consisted of a ten-seater bus equipped with a regular small car motor in addition to an electric motor. Nicknamed the "Centaur," this used the small-capacity internal combustion engine to drive a generator supplying current to the electric motor and to lead storage batteries for boosts on hills and for rapid acceleration.[23]

In the field of public urban transportation, most Soviet cities have a distinct environmental lead over their Western counterparts, which can be ascribed to shrewd foresightedness or to the advantages of backwardness, depending upon the reader's persuasion. In marked contrast with the noxious diesel bus now prevalent in the West, the Russians have continued to develop and to expand their trolley-bus and tram networks. Thus, during the current five-year plan period (1971-75), a further 2,956 kilometers of trolley-bus routes are to be opened and nearly 1,300 kilometers of additional tram lines are to be laid.[24] Many would maintain that trams tend to be too noisy and their inflexibility impedes other traffic excessively, but surely there is a strong case for the restoration of the trolley-bus in the West too. It is quiet, it emits no noxious fumes, and as for its restricted radius, an increasing number of cities are being obliged to set aside special lanes for buses anyway. Perhaps a straw in the wind was the city of Zurich's recent order for 11 trolley-buses from Hungary, at a cost of $81,000 each, in addition to 20 domestically produced models to replace its present diesel fleet.

About one-tenth of all air pollution in the United States and, presumably, a higher share in Western Europe and Japan, stems from domestic heating and refuse disposal. Here again, most Soviet cities are favored in that space or area heating and hot water supply is common and because most of the refuse is treated and incinerated centrally, mostly at combined heating/incinerator installations. The claim that "Soviet regional heating is first in the world" may well be justified.[25] Some 60 percent of the population of Soviet towns and settlements currently use centrally supplied hot and warm water for heating and washing purposes.[26] Another factor is the extensive use of natural gas: by 1975, it is said, two-thirds of the entire population will be using this clean and convenient fuel.[27] Its utilization has reportedly already reduced the level of air pollution in Moscow and Leningrad, for instance, by over 80 percent.[28]

Of course, the architecture of most Soviet cities and the relative youth of much of the urban housing stock assist in keeping down air pollution from heating and from refuse disposal. Many residents of Western cities prefer to have their own private family houses with gardens: this hinders the widespread adoption of regional heating or at least renders it prohibitively expensive. Their Soviet counterparts are not usually offered an alternative type of residence.

City and town planning in the USSR can proceed without many of the constraints known in the West, and the "master development plans" drawn up for several major cities for the forthcoming decades are indeed impressive. The general plan for Moscow, for instance, envisages the dismantling of about 200 polluting enterprises presently situated within city limits and flatly forbids the construction of any new offending plants.[29] It will be interesting to see if these 200 enterprises are in fact dismantled, for it has long been theoretically possible to declare an enterprise bankrupt and to close it down. But to this author's knowledge, this provision has rarely been employed, even though the financial status of many enterprises would have justified the action.

If the air over Soviet cities is relatively less burdened with emissions stemming from private cars, public transportation, domestic heating, and refuse disposal, it is nevertheless greatly degraded by truck exhausts and by industrial pollution, as may be confirmed by visitors to any of the major industrial cities. Thus, in Leningrad the brightness of daylight is said to be little more than half of that in surrounding areas because of industrial haze aggravated by the low-grade fuel used. The chemicalization drive of the 1960s has intensified air pollution; a prominent example is the notorious "foxtail" of chemical smoke from the Shchekino combine that spreads over the living areas to endanger the trees at nearby Iasnaia Poliana.[30] Certainly the urban authorities are aware of the problem and much is made of the establishment of special observation posts and automatic sensors for testing the purity of the air in large cities.[31]

A recent report suggests that the difficulties and expense incurred in neutralizing industrial emissions may have pressured researchers onto the wrong track. We learn that

> Soviet engineers maintain that chimneys taller than the Eiffel Tower are a reliable way of protecting cities from industrial pollution. Plans are being drawn up for the construction of a 320-meter-high chimney at Uglegorsk in the Donbass. Several such chimneys are to be built near Moscow, in the Baltic republics, and in the Ukraine. . . . Engineers are also working on a chimney that will emit smoke in a spiral into the upper atmosphere.[32]   TASS, Aug 28, 1971.

Clearly someone should remind those engineers of the old adage that "what goes up. . . ," as well as the host of more recent adages about the unity of man's environment.

Refuse Disposal

The horrifying vision, conjured up by the environmentalists, of Western man standing knee-deep in his own refuse may not become reality, but the battle against indestructible plastic bags, nonrecyclable cans, and the "no deposit, no return" bottle has only just been started. Soviet man has so far been spared this consequence of consumerism due to the backwardness of the various light industries such as, for example, all aspects of the food industry, and to a neglect of packaging technology. The bulk of the Soviet housewife's food purchases are still unwrapped: as late as 1970, a trade official of the Ukrainian SSR reported that only 6 percent of the flour, 3 percent of the sugar, and 5 percent of the butter in the republic's stores were sold in a packaged form;[33] it is hoped that the share of packaged foodstuffs in all-union sales by 1975 will rise to 60-65 percent.[34] Fortunately, Soviet newspapers are cheap and combustible in addition to their many other attributes. However, the belated adoption of self-service stores and, more recently, Western-style supermarkets (universamy) has obliged the foreign trade organizations to order packaging technology from Eastern and Western Europe, and there is no evidence to suggest that ecological considerations have played any role in their selection. But at least there is no sign as yet of the specifically "no deposit, no return" bottle on the Soviet market.

Noise Pollution

It is rather difficult to judge whether the eardrums of the average Soviet citizen are assaulted by more or by less noise than those of his Western counterpart. On the one hand, the urban resident hears less from automobiles, there seems to be less disturbance from jet over-flights, and the high-fidelity radio/gramophone/tape recorder—the ecstasy of the few and the bane of the multitude of apartment dwellers—is not present in great numbers yet, although recent letters to Izvestiia suggest that the age of the stereo is dawning.[35] Fortunately, also, with mass ownership of radios and television sets, the public loud-speaker system is not as ubiquitous as it once was. On the other hand, bus and truck drivers appear to exult in their noise-generating capability, and on various kinds of transportation—including jet airliners—the sound-proofing leaves much to be desired. The relatively few privately owned automobiles do make their presence heard: the practice of warming-up the motor periodically throughout winter nights has not yet been abandoned despite the advice of the motoring journal and the complaints of ground-floor residents. Western observers are frequently surprised at the high noise level tolerated in many Soviet factories and workshops. In the millions of new apartments built each year, the emphasis has clearly been placed on quantity rather than quality. Understandably, to have your own front door, after a lifetime of shared, communal apartments, is a "consumer good" far outweighing such minor considerations as doors and windows that do not fit or thin side walls, ceilings, and floors. But the current campaign to speed up the production and supply of ear plugs is indicative either of a growing noise level or a growing public awareness, or both.[36]

The detrimental effects of excessive noise upon the human organism have been fully acknowledged by the Soviet authorities. A National Anti-Noise Committee was set up in September 1969. Among its achievements have been the design and production of a thicker Keramzit concrete block for prefabricated houses and the enforcement of regulations keeping the noise level in residential districts of Lithuanian cities below the established norms of 35 decibels by day and 30 decibels by night.[37] Another guardian of the eardrums, the chief public health officer in Moscow, has promised that trolley-buses will be retained there because of their quietness of operation. He also emphasized that residential blocks in the city would in the future be situated away from thoroughfares and behind stores and public buildings.[38]

In a more controversial sphere, official spokesmen have categorically denied that the TU-144 supersonic transport will create any unacceptable noise level for those living near its airports, or for those in the path of its sonic boom.[39]

Conservation

The record of the Soviet government on protecting and preserving the natural resources of its vast territory has been very mixed. To be sure, successive administrations have passed legislation dealing with various aspects of conservation, from Lenin in 1919, to Khrushchev's Party Program of 1961, to the many decrees of the present administration.[40] But the early, heroic settler's attitude toward nature—that it is there to be fought with and overcome—still permeates the thinking of many in authority and the spirit of economic decisions. Only lately have there appeared signs of a grudging comprehension that man must work with and not against nature, and only comparatively recently have steps been taken to rectify damage done in the past and to conserve what has been left unravaged. Thus, one broadcast spoke of "105 projects where Soviet scientists have artificially restored the balance of nature, including the breeding of species that had become almost extinct."[41]

Probably the greatest single act of degradation committed on Soviet soil was Khrushchev's Virgin Lands campaign. Faced with the dire necessity of raising grain output in the shortest possible time, he chose the correct initial course in that given context. His crowning error was to persist with his pet project; instead of switching the investment resources and manpower increasingly back to the traditional grain areas after his first successes in Kazakhstan and Siberia, he subjected these marginal soils to monoculture year after year, disregarding the experts' advice as to fallowing and agrotechnology. The results were vividly chronicled in the Soviet press even before Khrushchev's departure from office: the topsoil from millions of hectares blew away, the resulting dust storms turning day into night and obliging truck drivers to keep their headlights on at all times. After a few years, the exhausted soil often returned yields that barely covered the seeding norms.

In the early part of the last decade, it was primarily the literary journals that aroused public awareness of the callous lack of foresight and concern for other consumers and interests evidenced by ministries, administrations, and local officials,[42] but of late the cause has been taken up by many of the nation's daily newspapers. Sometimes the motivation has stemmed purely from environmental considerations: an example of this was the eloquent protest against the decimation of bears by privileged Intourist-sponsored, hard-currency-paying hunters.[43] On other occasions, conservation may have been used as a stick with which to beat a political opponent; for instance, there was the revelation that Minister of Agriculture, Vladimir V. Matskevich shot deer in a game preserve.[44] Was this a case of unsullied altruism or was it perhaps sponsored by his chief opponent, the head of the government of the Russian Republic, Gennadii I. Voronov?

Yet, despite a growing exposure by the media, flagrant offenses against conservation interests continue to occur and, indeed, are built into current medium-term and perspective plans. For instance, the hills above Petropavlovsk have been denuded of their forests for lumber, leaving the city unprotected against torrents of muddy water in the summer and avalanches in the winter.[45] The scale of open-pit or open-cast mining is being vastly extended: by 1975, over 30 percent of all coal mined, that is, some 208 million tons per year, will be extracted by the open-pit method.[46] It has been argued that most of this will take place in remote regions, far from human habitation, but even the Soviet Union is not so vast that the laying bare like gaping sores of perhaps millions of hectares can be thus dismissed.

On the international scene, Soviet representatives have been vocally most active in the cause of conservation. In 1964 the USSR and Norway agreed to suspend all hunting of the Greenland seal for five years.[47] In 1965 Soviet spokesmen publicly criticized the admittedly crude U.S. practice of hunting polar bear from aircraft.[48] In 1966 the USSR announced that it had banned the catching and killing of dolphin,[49] and between 1967 and 1969 it joined with the United States in two pacts for the conservation of crabs and certain fish in the Pacific and the Atlantic.[50] These were all very laudable incentives, but they represented areas of marginal economic concern for the USSR. In marked contrast has been the Soviet record on whaling. Whales are a valuable source of meat and by-products for the USSR which, together with Japan, accounted for some 35,000 of the 42,000 whales taken in 1970.[51] The Soviet whaling fleet is modern, impressively equipped, and represents a very substantial investment. However, whales have now been so overhunted that they face extinction. In 1971 the United States halted all whaling for that year. Before the 1971 season started, negotiators from the USSR, Japan, and Norway initialed an agreement to the effect that international observers would accompany each ship to ensure that the number of whales taken did not exceed the quotas set by the International Whaling Commission and to guard against the killing of protected species. After the agreement was initialed, the Soviet whaling fleets slipped away from their home port of Vladivostok before the international observers could board.[52] Unabashed, First Deputy Premier Mazurov lectured the assembled delegates to the Northeast Atlantic Fisheries Commission session in Moscow a few weeks later on the necessity for conserving the wealth of the oceans.[53]

ECONOMICS OF POLLUTION

In attempting to assess and to allocate the social and private costs of a new industrial plant, a technological process, or a water management project, Soviet planners face many of the problems familiar to their Western counterparts plus additional ones attributable to the peculiarities of the Soviet economic system. What value is to be put on the disfigurement of a beautiful landscape, the raising of the noise level in a community, or the defiling of its atmosphere? Do the additional job opportunities created by a new plant offset the costs imposed? As in the West, infrastructural overheads are economized on when a new polluting enterprise is "planted" in an existing metropolis:[54] where do the ensuing stink, noise, and dirt appear in the balance sheets? Such questions arise for Western and Soviet planners alike.

Occasionally the economics of environmental disruption in a given instance are reduced to a simple trade-off formula. Thus, biologists and Baku residents may argue that restoring the sturgeon catch from the Caspian Sea to its former level would be financially more rewarding than continuing oil extraction anywhere near the shore.[55] Erosion control measures costing X million rubles could raise the annual grain harvest by Y million tons; open-pit mining yields XX millions of tons of coal, but reduces the area of arable land by YY thousand hectares, and so on. But even such apparently clear-cut equations are misleading.

To date, the Soviet authorities have not deemed it necessary to apply a specific scale of charges for pollution that would oblige plant economists to compare the economic sanctions against a new piece of technology with the gain from projected output. For instance, the U.S. Environmental Protection Agency is reportedly considering a tax on sulfur of initially 1¢ per pound of emission, later to rise to 10¢ and, in the worst areas, to 15¢ per pound on the assumption that somewhere along that time scale it will pay offending plants to install the necessary pollution control device.[56] Almost the opposite occurs in the USSR: in many instances the set fine for pollution goes into the coffers of the local authorities and is used to finance a new club or garden.[57] The community thus acquires a vested interest in the continuation of environmental disruption.

The few Soviet economists who have been courageous enough to try to quantify the costs—in terms of economic growth foregone—of eliminating pollution or of reducing it to tolerable levels have tended to understate the expenditures required. The consensus has put the cost at a few percentage points at most to be deducted from the annual growth rate.[58] But an Organization for Economic Cooperation and Development (OECD) study on U.S. pollution estimated that to bring about a noticeable slowing down in pollution would require the reallocation of some 2 percent of the GNP, to stop the progress of deterioration would take 4 percent, while to effect any tangible repair of the environment would require the reallocation of 8-12 percent of the GNP. However, since any ecological improvement measures would result in a slowing down of overall economic growth, the commitment in relative terms would be greater.[59] Of course, the economic effect of environmental quality control on current national accounting aggregates depends upon how related activities are recorded. As a recent European Economic Community (EEC) study points out, by Western measures "the GNP is bigger when pollution occurs and is expensively remedied than when it does not occur at all."[60] (Presumably pollution control measures, being nonproductive, would not figure in the Soviet concept of national income.) For the next decade or so, the Soviet economy is expected to rank at best somewhere in the middle of the medium growth league, that is, on a par with the EEC states and only slightly ahead of the United States. There is no evidence to suggest that the Soviet leadership would be prepared to jeopardize this already uninspiring perspective on environmental grounds.

19

The principal ideological yoke that the environmental lobby in the USSR must still bear is the Marxian prejudice against charging for natural resources. Of course, the more progressive economists have long protested the irrationality of this dogma. Some five years ago, Academician Fedorenko wrote: "The idea that natural resources are free of cost has become the yesteryear of Soviet economic science, although it has been overcome in economic theory only."[61] During the 1970 discussion prior to the adoption of the "Bases of Water Legislation of the USSR and the Union Republics," there were repeated demands for the institution of a charge for water. But Article 15 of the adopted "bases" propounds the continued free use of water. The explanation offered is: "Were water-use payment to be exacted in every case, it would create problems of a complexity beyond the powers of either economic science or practice."[62] Yet the argument made by Academician N. Melnikov, the president of the Commission for Studying Productive Forces and Natural Resources, remains irrefutable. While water is supplied free of charge, the enterprise director has no incentive—other than moral—to economize in its use. If a charge were made of, say, ten kopeks per cubic meter of pure water, then an enterprise would find it advantageous to recycle this at a cost of, say, six kopeks per cubic meter.[63] The same is true for other natural resources, with predictable consequences. As Aleksander Birman explained:

> At present, the enterprises of the mining industry are receiving the riches of the depths free of charge, although society spends sizable funds on their discovery and prospecting. Since the enterprises get the depths of the earth free of charge, they have no incentives to exploit the deposits to the end, since costs increase and profitability declines as depth into the shaft-mine or oil-stratum increases. Therefore, up to 25 percent of existing reserves remain in the deposits.

He also emphasized the irrationality of enterprises receiving water free of charge while the state spends billions of rubles on the creation of water reserves.[64]

ORGANIZATION OF MEASURES TO COUNTER ENVIRONMENTAL DISRUPTION

Several Western governments have belatedly recognized the need for a national body to watch over the country's environment. It must be capable of collecting and collating the necessary information, of formulating a basic economic framework that can assess environmental problems in terms of costs and benefits, and of initiating and ensuring the implementation of preventive and corrective legislation. Institutions like the U.S. Environmental Protection Agency (EPA) and the U.K. Standing Royal Commission of Environmental Pollution have been set up and are gaining in authority and effectiveness, although few would deny that their work has only just begun. The EPA, for instance, faces a herculean task in its confrontation with the extractive industries, traditionally favored by the mining laws.

Environmentalists in the USSR have also long been campaigning for an all-union body with meaningful powers over ministries and local authorities. According to one critic, pollution research is being conducted by small and scattered groups of scientists who have inadequate experimental facilities and no rapid methods of even determining the level of pollution. He called for the organization of an "integrated pollution control service."[65] Others have proposed the creation of a "scientific coordinating institution for exploiting nature,"[66] although these missed the point that the concept of "the exploitation of nature" touches on the work of virtually every branch and sector throughout the economy. A third critic was more to the point in suggesting the establishment of a State Committee for the Preservation of Nature.[67] (East Germany has recently gone one step further in setting up a Ministry of the Environment and Water Management.)[68]

The lack of an overall supervisory and executive body is illustrated by the conflict of many ambitious projects affecting the Aral and Caspian seas. Of these an observer remarked: "Each department can defend its own interests and put forward its own reasoning. But there is not one agency among them that is wholly responsible for the conservation and augmentation of our national resources."[69] On the same theme, an Azerbaidzhani official noted that his republic had four organs concerned with the conservation of the Caspian Sea, but there were three other Soviet republics bordering the sea, in addition to Iran, and no effective coordination had been attained.[70]

At present, the protection of the environment in the USSR would appear to rest in the hands of a plethora of uncoordinated, largely advisory, bodies plus a large but seemingly impotent network of concerned citizens.[71] The former are to be found at all levels and in all regions, such as one rejoicing in the title of the "Special Scientific Council on the Problems of Protecting the Air from Pollution from Harmful Substances of the USSR Council of Ministers' State Committee for Science and Engineering,"[72] an All-Union Research Institute of Water Conservation,[73] the USSR Academy of Sciences' Commission for the Study of Productive Forces and Natural Resources,[74] and a Standing Commission of the Azerbaidzhani SSR

Supreme Soviet for the Protection of Nature.[75] Concerned citizens, of whom there are said to be over 16 million, may join "public nature protection committees" under the auspices of the Central Council of the All-Russian Nature Protection Society.[76] These are all very impressive, but they hardly add up to a universal and effective system for defining, detecting, and deterring offenses against the environment.

One critic put his finger on the crux of the matter when he pointed out that there has not been an all-union agency, responsible for preserving the environment that was not itself involved in the exploitation of natural resources, since A. V. Lunacharskii's Main Science Administration of the People's Commissariat of Education in the 1920s.[77] Clearly, when a senior official of the USSR Ministry of Land Improvement and Water Management is made the president of the RSFSR* Society for the Protection of Nature,[78] or when the Ministry of the Fishing Industry is held responsible for the conservation of fish stocks, "the poacher is appointed gamekeeper" and a conflict of interests is inevitable. When production targets clash with conservation imperatives, it is not difficult to guess the outcome in the Soviet context.

One of the major weaknesses of the Soviet ministerial system, and one that Khrushchev tried to remedy in his impulsive manner, is its "departmentalism." Usually this pejorative term implies that a ministry consciously rides roughshod over the interests of others, but it also applies to situations where the ministry may just be ignorant of other interests. The USSR Ministry of Power and Electrification is a case in point. On one of the rare occasions when USSR ministers traded personal recriminations in public, Matskevich accused Neporozhny at the 24th Party Congress of needlessly ordering the flooding of thousands of hectares of good arable land for his power stations.[79] In this instance, the power minister had presumably known of, and consciously chose to ignore, agricultural interests. But on previous occasions, such as the flooding of enormous reserves of petroleum, gas, lignites, and nonferrous metals by the construction of the Nizhne-Obskaia GES and the relocation of 110 kilometers of the Taishet-Lena railroad track because of the Bratskaia GES,[80] the crossing of the wires may well have been due to lack of information stemming from the vertical structure of the Soviet economy.

LEGISLATION

Russian and Soviet legal history abounds with measures providing for the conservation of natural resources and prohibiting

---

*Russian Socialist Federated Soviet Republics.

the pollution of water and air, the wasteful use of land, etc. One of
the simplest and most effective of the early laws was Peter the Great's
Ukase of 1720, which prescribed capital punishment, flogging, branding,
or imprisonment for any who presumed to fell trees in the royal forests. In 1913 a draft law was submitted to the Duma "on protecting
the purity of air, water, and soil."[81] Much is being made nowadays
of the special "shock-work" decree on the water supply, which Lenin
signed shortly after the revolution,[82] but this was presumably more
concerned with simple sanitation. The first Soviet nature preserve
was set up in 1919,[83] and even during the chaotic conditions of war
communism, the Supreme Economic Council (VSNKh) found time to
pass a decree "On the Setting Up of a Main Administration of Water
Management and Land Improvement."[84] Since those heroic days,
enough legislation of this nature has been enacted to fill at least one
sizable volume, namely Okhrana prirody, edited by V. M. Blinov
(Iuridicheskaia literatura, Moscow, 1971).

The supreme criterion of any legislation is its effectiveness.
In the field of environmental legislation, the Soviet record is sadly
similar to that of many Western countries. After all, Germany has
had a law protecting the environment on the books since 1930, but the
Rhine is a moving testament to that law's impotence. In the published
Soviet debate, it is occasionally admitted that particular items of
environmental legislation have been wholly ineffective. Referring to
a 1964 joint oblast party committee and executive committee decree
on combating pollution in the local river, a critic in 1970 concluded:
"Alas, not a single point of this decree has been implemented."[85]
But this was unusually frank, and more frequently the impotence of
earlier legislation is merely implied by the passing of new laws
covering much the same ground. Thus many of the provisions of the
water conservation decree of 1960 were repeated in the water legislation of 1970.[86] Similarly, the issuing of a second decree in less
than three years on the conservation of Lake Baikal could be hailed
as evidence of undiminishing concern.[87] It also demonstrated the
ineffectiveness of the earlier measure.

In general, Soviet environmental legislation has lacked the
specificity of recent Western measures. For instance, the 1970 Clean
Air Act in the United States ordered the Environmental Protection
Agency to stipulate a set of national air quality standards. This it
has done with admirable exactitude.[88] Such rigorous standards may
not stick—Detroit is apparently conducting an effective spoiling
action—but their specificity is in welcome contrast to the largely
hortatory provisions of, say, the decrees on the Caspian Sea and Lake
Baikal, if the texts published in the press reflect the decrees faith-
fully.[89]

DETERRENCE

The deterrents for despoilers employed by Peter the Great have long since been discontinued by his successors, although it might be noted that the Soviet Union is one of the few developed nations that still applies capital punishment for economic crimes. Mercifully, the pollution of water and air does not come under this category.

The Soviet authorities recognized, before most Western administrations, that moral strictures and fines do not always constitute adequate deterrents against the disruption of the environment. Most of the environmental offenses dealt with in the early 1960s were punished with the withdrawal of bonuses, monetary fines, or, in the more flagrant cases, with the dismissal of the culprit. The one-shot confiscation of a bonus or the levying of a nonrecurrent fine provided no lasting constraint, however, and the only meaningful long-term deterrent was shown to be the continuing loss of bonus payments or a repeated fine until the offending plant or equipment had been rectified.[90] The deterrent effect of fines has further been reduced by two principal factors. In the first place, their scale is still generally insignificant. The maximum fine for poaching beluga sturgeon for some unexplained reason was recently halved to 100 rubles.[91] The same fine is usually applied for the pollution of land with waste or sewage.[92] However, more substantial fines are provided for in the legislation,[93] and they are occasionally levied. Thus the director and the chief engineer of an Estonian cellulose plant were fined 2,000 rubles for polluting a fish hatchery.[94] The second shortcoming of monetary sanctions has been that the fines were, more often than not, paid out of the enterprise's working capital and not from the offending official's pocket. This weakness now seems to be generally recognized.

In the West, the actual imprisonment of environmental offenders is still so rare that it makes the headlines; suspended sentences are more common. Soviet courts, on the other hand, have been awarding jail sentences rather more frequently, and especially during the past two years. For example, the assistant chief engineer of a sugar refinery was given one year of corrective labor for polluting a local river with untreated waste,[95] while the captain of a dredge was hammered with a five-year jail sentence for poaching sturgeon.[96] The wider use of jail sentences for environmental offenses may be attributed partly to legislation with more teeth and also to the intensified emphasis upon labor discipline discernible since the December 1969 Central Committee plenum.

Existing legislation empowers the authorities to shut down polluting enterprises,[97] although the present author has found only a few instances where this sanction has actually been applied.[98] It

24

should, of course, be noted that much the same is true of Western environmental legislation. The temporary closing of 23 plants in Birmingham, Alabama, in November 1971 represented one of the first tastes of really firm action by the Environmental Protection Agency and a possible portent of more federal intervention in the future.

PUBLIC ATTITUDES

It is probably true to say that, in their attitude toward the conservation of natural resources, most ordinary Russians are at best apathetic. The feeling that "the earth is big enough to take care of everyone"[99] has been inculcated in the Soviet citizen from his first days at school and only recently has a more discerning line been adopted by the mass media. Endless films have featured endless vistas of the taiga, the vastness of the steppe, and the expansiveness of the rivers. Heroic engineers, prospectors, and tractor drivers are pictured erecting huge dams, blasting for oil, and ripping up virgin soil. "From time immemorial," it was observed, "the inhabitants of Russia have been accustomed to thinking that the expanse of their homeland is infinite and that its riches are inexhaustible."[100] The same author noted that Soviet students have traditionally been imbued with an "aggressive attitude toward nature," with emphasis put upon the "struggle with nature, the taming of nature, and the transformation of nature."[101]

Similarly, it would be misleading to picture more than a relative handful of Soviet citizens as ecological activists on, say, the questions of industrial pollution, automobile exhausts, or nonrecyclable refuse. On viewing or smelling smoke from factory chimneys, the feeling of most Russians probably varies from the pride of a developing nation in its new-found industrial muscle to the tolerant acceptance encountered in the north of England that "where there's muck, there's money." For an aspiring motorist who has never possessed a car of his own, who has saved the equivalent of four years' take-home pay for a modest sedan, and whose name has been on the waiting list for three or more years, any admonition that his new car will add "X" grams of pollutants each year to the atmosphere will fall on deaf ears. For a housewife accustomed to collecting the family's milk in a can and to having her sugar wrapped in Vechernaia Moskva, the disposal of millions of bottles or the indestructibility of plastic packaging are marginal considerations.

To be sure, the inhabitants of Baku or Riga resent the film of oil that covers them when they emerge from a dip in the sea, but they would not dream of manifesting their resentment with acts of "ecotage" directed against the offending derricks or tankers. Instead, this anger is channeled into the docile "public nature protection committees," or oily bathers are asked to telephone their complaints to, say, the Riga Zonal Water Inspection of the Ministry of Land Improvement and Water Resources of the Latvian SSR.[102] As yet there have formed only the nuclei of pressure groups that could one day translate the vital interests of the bulk of the population into action. Effectively to combat the environmental by-products of industrialization and modernization, more members of Soviet society need unimpeded and unfiltered access to the research and experience of other societies that have trodden the same path. But this free access is denied to all but a handful of selected and/or preeminent scientists and officials. The eloquence of Academician Sakharov is moving and his logic persuasive, but the lesson of other societies in the struggle for ecological survival is surely that the driving force generally comes from below and may not effectively be imposed from above.

MEDIA TREATMENT

The inner thoughts of Chairman Kosygin, General Secretary Brezhnev, and other Soviet leaders on the subject of environmental disruption have not been disclosed, apart from a few perfunctory phrases at party congresses and similar gatherings. At the November 1971 session of the USSR Supreme Soviet, for instance, Premier Kosygin devoted merely one sentence to this topic.[103] The only indications of leadership attitudes are thus to be derived from the imperfect and cloudy mirror of the media in general and the press in particular. These are at best tenuous, for while most newspapers and journals follow the "party line" for most of the time, Glavlit, the Main Administration for Literary Affairs and Publishing, is not ubiquitously effective; moreover, the airing of an issue may sometimes betoken the existence of a disagreement within the leadership.

Until comparatively recently, pollution—like unemployment, drunkenness, crime, and plane crashes—was officially treated as a by-product and concomitant of capitalism exclusively. It apparently did not exist at all under socialism or was present only to an insignificant degree; in the latter event, it was not discussed much in polite circles. A typical pronouncement went: "The problem of air pollution in a number of capitalist countries has reached the level of a national disaster. . . . Of course, our state would never permit such a thing."[104] A more realistic and less righteous tone is generally adopted nowadays, although as late as 1971 such passages appeared as: "This attitude toward water resources [i.e., reckless contamination] is still characteristic of capitalist nations,"[105] and "as ecology veteran Eugene Odum once remarked sadly, 'The American credo is to enrich oneself today and to let tomorrow take care of itself.' Such an approach—'the extensive plundering of the future'—we reject."[106] TASS promptly reports a burst pipe in an offshore oil well belonging to the Iminico company off Teheran,[107] but a similar break off Baku may become known only months or years later and then may have to be gleaned indirectly from a published exchange.

The image of environmental disruption in the USSR, which is served up for external consumption, differs from the increasingly frank domestic debate. One Soviet delegate to the 1970 U.N. Economic Commission for Europe symposium on environmental problems in Prague contrasted the "purity" of the Moscow River with the filth of the Potomac, the Thames, and the Seine,[108] although, as we have seen, "purity" is hardly the term to use once the river reaches the metropolis. Another Soviet spokesman at the same symposium claimed that the intensity of pollution in the USSR was declining, despite a considerable growth of industrial capacity. He was cited as adding, gratuitously and ambiguously, "The problems of the environment are closely linked with the planned development of the national economy and the location of factories."[109]

The Soviet authorities censored a Russian-language edition of the July 1971 issue of the UNESCO magazine, Courier, deleting a report that cited some of Marshall Goldman's findings on pollution in the Soviet Union and carried an unobjectionable article by the president of the Ukrainian Nature Protection Committee.[110] The examples employed had all appeared in the Soviet press, but presumably the implied parallels were considered to be invidious.

On another international issue, Vladimir Kunin, the Soviet delegate to a Finnish conference on the environment, unwisely confided to a Western journalist that some of his colleagues were concerned about the possible effects of SST (supersonic transport) exhaust on the stratospheric ozone shield and thus on the amount of harsh radiation reaching the earth.[111] This disclosure conflicted sharply with the official position adopted by A. F. Aksionov only a few days earlier, namely that no adverse ecological effects from SST operations were anticipated.[112] But Western government spokesmen have not always been completely frank on this and other environmental problems either.

If the exposure—within limits—of pollution and other environmental offenses in the Soviet press now has the evident blessing of the authorities, any parallel with Western movements is refuted. The ecology campaign in the United States is deemed to be merely a new vehicle for "social demagogy" and is exploited as a "shock absorber" for oppositional sentiments.[113]

As has been mentioned, the most eloquent defenders of the environment have been, and remain, the literary journals such as Novy Mir, Nash sovremennik, and Literaturnaia gazeta, with stout support from the youth newspaper Komsomolskaia pravda. However, even in the early 1960s, it was not unknown for a battery of national, local, and specialist journals to bring their combined weight to bear on an environmental issue. For instance, when the climate of Kislovodsk was threatened by the removal of a large chunk of its mountain shield by the depredations of the lime-works at Podkumka, the campaign to save the resort was joined by Izvestiia, Trud, Literaturnaia gazeta, Sovetskaia Rossiia, Kavkazskaia zdravnitsa, and Meditsinskii rabotnik.

The environmental lobby in the USSR must have been disappointed by the much-bruited and long-awaited publication of the journal Ekologiia, which was launched in 1970. Not only did this have a very limited edition—so far not exceeding 1,275 copies—but, to judge from the issues that have reached the West, it is exceedingly timid in its treatment and has confined itself largely to problems of general biology. A stauncher champion of ecological issues has proved to be the journal Priroda.

As with other officially approved causes, suitable benedictions from the founding fathers have been unearthed, although these have tended to be less than specific. Thus Marx is cited as writing: "Cultures that develop randomly, and are not directed consciously, leave deserts in their wake";114 while Engels' contribution is: "We will not, however, flatter ourselves too much with our victories over nature. For every victory, it takes its vengeance upon us."115

It is reassuring to note that growing attention is being devoted to environmental disruption and to the necessary countermeasures in samizdat (self-publication) writings. If the authorities were to decide, for any reason, to play down or completely to suppress the discussion of an ecological issue in the usual media, then samizdat would provide an alternative channel for dissemination. Already, well-informed and cogent discussions on ecological and conservation themes have appeared in such diverse samizdat publications as Sakharov's Progress, Coexistence and Intellectual Freedom (Arkhiv Samizdata 200*), Sakharov, Turchin, and Medvedev, Appeal to the Soviet Leadership (AS 368), Veche, No. 1 (AS 1013), and Maksimov, Seven Days of Creation (AS 1047). And, according to Khronika tekushchikh sobitii, No. 22, a detailed study of pollution in the USSR will be forthcoming in Veche, No. 3.

---

*These numbers refer to the Archive of Samizdat located at Radio Liberty in Munich.

INTERNATIONAL COOPERATION

As noted before, the USSR has already entered into several international agreements on the conservation of various species of animals and fish. In the field of pollution also, Soviet spokesmen have conceded the unity of the world's environment.

In his famous memorandum, Academician Sakharov called for international cooperation in combating pollution; otherwise the USSR would poison the U.S. environment with its waste and vice-versa (AS 200). This was echoed by two environmentalists in a conventional publication:

> The protection of the natural environment is acquiring an increasingly international character. Indeed, rivers, ocean currents, and air masses move according to the laws of nature, without acknowledging state boundaries. Water or atmospheric air that has been polluted in one place can have a pernicious effect on the biosphere in other parts of the planet. Therefore, in recent years, the U.N. has shown increased interest in the problems of conservation.[116]

In 1971, Premier Kosygin devoted most of an interview with four U.S. industrialists to the topic of the environment, urging that the United States and the USSR cooperate in a major fight against pollution and during President Nixon's visit to the Soviet Union in May 1972, the two nations initialed a treaty aimed at reducing environmental deterioration.[117]

Soviet representatives have played a prominent role at international gatherings on conservation and environmental problems. For instance, the 1966 International Wild Life Conference in Lucerne was assured of full Soviet support,[118] while a Soviet scientist presided over a meeting of the International Union for the Conservation of Nature and Natural Resources.[119] An agricultural official of West Germany has recently praised the Soviet initiative in initiating a conference aimed at regulating fishing quotas in the Northeast Atlantic.[120]

In the south, Soviet and Iranian authorities have agreed to consult on joint action to clean up the Caspian Sea,[121] while in the north, the pollution of the Baltic Sea was strongly deplored by a TASS statement. This implicitly disclaimed any Soviet responsibility and named Sweden, Denmark, and "other Baltic nations" as the culprits. Nevertheless, it called for "a further expansion of joint research work in the Baltic with the participation of all the Baltic countries under agreed international programs."[122] However, a later domestic broadcast suggested that the USSR is contributing more than its mite to the current sad state of the Baltic. A departmental head of the Lithuanian Academy of Sciences claimed that if industrial effluent were disposed of at a depth of 22 meters, at about 6 kilometers from the shore and if it did not contain more than 1 mg of ether-soluble substances per liter, then "no pollution of the Baltic would occur."[123]

On the whole then, the Soviet record of international consultation and collaboration during the past decade has been as impressive as that of most nations, apart from the flagrant violation of the whaling agreement.

A BALANCE SHEET

In common with all other industrialized nations, the USSR has environmental problems: such differences as do exist are largely those of degree and stage of development. Thus the Soviet population density is well below those prevailing in the United States, Western Europe, and Japan, but most of it lives in a relatively small portion of the Russian land mass. The harmful side effects of chemical fertilizers, herbicides, and pesticides may not have been manifested on the scale experienced, say, in Western Europe, but they may do so once the USSR reaches the intensity of application now practiced in those countries. The 1.5 million privately owned automobiles clearly do not emit as much pollution as the 85 million in the United States, but the Soviet automobile stock will have quadrupled by the end of this decade and most of it will be concentrated in the urban areas. In the field of packaging for foodstuffs and other consumer goods, the USSR Ministry of Light Industry and the USSR Ministry of Trade are importing or emulating Western technology: unless they learn from the West's mistakes and profit from the Western debate, they will be confronted with similar problems of refuse disposal. The Soviet housewife is discovering the convenience of household detergent: presumably Russian rivers will also foam in due course.

Many other similarities affecting the environment may be discerned. If, in order to increase output, reduce prime costs, and enhance profitability, the Soviet enterprise director must further pollute the air, water, or land, he will probably do so—as will his Western counterpart—in the absence of adequate economic or penal sanctions. He is thereby transferring the "social costs" of his plant's operation on to an unspecific generality that is usually unprotected or inarticulate. The manufacturer in the USSR, as in most Western nations, still does not bear the burden of proof of showing that society will not be harmed by his product or production process. The social costs, as opposed to the private costs, of a new technology are difficult to quantify and to allocate. Polluting plants will be sited in already-inhabited localities to save on infrastructural costs, and their noxious effluents will be discharged untreated into already-overflowing public sewage systems, unless prevented from so doing by an effective watchdog organization.

Where, then, do the Soviet experiences and attitudes differ? From the narrow point of view of protecting the environment, the Soviet political, economic, and social structure offers certain distinct advantages. For example, the greater de facto degree of authority enjoyed by the union over the republics, or by the state over the individual, the easier it is to enact and to implement legislation against pollution. States' rights or private property interests present no obstacles there. Where planners' preferences reign over consumers' choices, then it is much easier, say, to plan city layouts, build high-ways, and site airports. Those citizens who wish to live in urban areas are generally obliged to reside in large apartment buildings that are regionally heated; this undoubtedly reduces pollution from heating and from refuse disposal. If the decision maker in Moscow feels, for instance, that the "snowmobile" or "skidoo" is a superfluous abomination, which fouls the pure air of the mountains and shatters its peace—and here he would have the present author's full support—then the Soviet citizen will simply be given no opportunity of purchasing or operating such a machine. Or, to relate to an as yet only latent threat in the West, it is unlikely that a Soviet entrepreneur will be able to exploit the market for some specific pollution control device and make a fortune out of it, to the public detriment.

On the other side of the coin, many aspects of the Soviet economic system militate against environmental interests. The absence of a charge, or the inadequacy of charges, for natural resources leads to their unbridled use or profligate waste. As two leading Soviet con-servationists put it: "The costs of conserving and restoring these resources do not enter as a component part into the enterprise's production costs."124

Rapid economic growth has historically been a primary objec-tive of Soviet policy and its maintenance has helped to legitimize party control over planning and management. However, by any meas-ures, Soviet economic growth slowed during the 1960s and is not expected to stand out in the international growth stakes during the foreseeable future. Any additional impediments to growth, such as those attributable to environmental considerations, are hardly likely to be accepted by the Soviet leadership. The prime criterion of an enterprise's performance and the main yardstick for awarding premia remains the growth of gross output: when this conflicts with ecological measures, the outcome is predictable. Material incentives for pre-serving the environment are marginal.

One of the principal legacies of the "Stalinist" model has been the unevenness of Soviet economic development. Key sectors were accorded priority and recorded impressive growth, while the remaining sectors were treated as residual claimants. Together with consumer services, trade, and other "nonproductive" spheres, environmental protection has been one of these residual claimants. Such ordering of priorities has led to the allocation of insufficient resources for research and development on pollution control technology, little capacity devoted to its production, and a low status accorded to its specialists. The citizen has been relegated to second place not only in the mix of industrial output but also as a consumer of the benefits of the natural environment.

Despite the existence of a plethora of organizations and societies concerned with protecting the environment, in many instances the ultimate responsibility for preserving natural resources lies with the agency charged with exploiting them. A one-sided conflict of interests is thus inevitable. Moreover, as we have seen, the vertical structure of the economy may impede the flow of information concerning environmental disruption and reduce the effectiveness of countermeasures.

Control of the media has meant that the officials, executives, and technicians concerned, as well as the broad public, are less than fully informed as to the successes and shortcomings of other nations in curbing environmental disruption. It has also meant that only "approved" debates are aired: thus the misgivings of certain scientists regarding the possible damage caused to the stratospheric ozone shield by SST flights would not be disseminated. Samizdat offers a supplementary and alternative channel of propagation, but as yet its radius is limited and other issues are paramount.

Nevertheless, the Soviet authorities are by now aware of the fundamental issues involved and the choices to be made if environmental disruption is to be checked and/or corrected. Official pronouncements and approved debates acknowledge the urgency of the problem. As yet, no case has arisen that would unambivalently identify the leadership's true priorities in this respect.

NOTES

1. See, for instance, Marshall Goldman, "The Convergence of Environmental Disruption," Science, October 2, 1970, pp. 37-42, "Externalities and the Race for Economic Growth in the USSR: Will the Environment Ever Win?" ASTE Bulletin, Spring 1971, pp. 19-27, "The Pollution of Lake Baikal," The New Yorker, June 19, 1971, pp. 58-66; and Marshall Goldman and David E. Powell, "The Social Costs of Modernization: Ecological Problems in the USSR," World Politics, July 1971, pp. 618-34.

2. International Herald Tribune, April 27, 1970.

3. Za rulem, no. 1 (1972), p. 38. Apart from privately owned automobiles, of course, official cars, taxis, trucks, etc., contribute to air pollution. The estimated current stock of all motor vehicles is roughly 7 million in the USSR, compared with about 110 million in the United States.

4. See NATO, Prospects for Soviet Economic Growth in the 1970's (Brussels, 1971).

5. Izvestiia Akademii Nauk SSSR: Seriia Ekonomicheskaia, no. 2 (1971), pp. 35-47.

6. See this author's "Soviet Agriculture in the 1970's," Studies on the Soviet Union, no. 3 (1971), pp. 1-45.

7. See, for instance, Pravda, October 7, 1968; Komsomolskaia pravda, August 26, 1970; The Observer, July 18, 1971.

8. Pravda, December 27, 1971.

9. Khimiia i zhizn, no. 1 (1970), p. 51.

10. Literaturnaia gazeta, no. 10 (1971), p. 11.

11. Nedelia, no. 12 (1968), p. 14.

12. Nauka i tekhnika, no. 9 (1969), p. 2.

13. Gorodskoe khoziaistvo Moskvy, no. 6 (1971), p. 42.

14. Moskovskaia pravda, September 1, 1968.

15. Vestnik statistiki, no. 3 (1970), p. 11.

16. See, for instance, Sovetskaia Latviia, December 14, 1969.

17. TASS, January 16, 1972.

18. Sotsialisticheskaia industriia, December 24, 1971.

19. Narodnoe khoziaistvo Kazakhstana, no. 2 (1968), p. 83.

20. TASS, October 21, 1970.

21. Radio Moscow-I, 1600 GMT, January 17, 1972.

22. For a review of the latest results, see Za rulem, no. 12 (1971), p. 11.

23. TASS, August 23, 1969.

24. Ekonomicheskaia gazeta, no. 4 (1972), p. 17.

25. Teploenergetika, no. 7 (1971), p. 35.

26. TASS, January 13, 1972.

27. TASS, January 21, 1972.

28. Gazovaia promyshlennost', no. 4 (1970), p. 37.

29. TASS, October 21, 1970; cf. Moscow News, July 24, 1971 and the International Herald Tribune, January 19, 1972.

30. Komsomolskaia pravda, April 3, 1968.

31. Thus Radio Kiev in Ukrainian for Abroad, 1800 GMT, October 8, 1970 and Radio Kiev, 1615 GMT, September 16, 1971.

32. TASS, August 28, 1971.

33. Pravda, March 5, 1970.

34. Izvestiia, November 18, 1971.

35. Izvestiia, January 25, 1972.

36. Ibid.

37. TASS, September 8, 1970.

38. Radio Moscow-2, 1100 GMT, September 13, 1971.

39. International Herald Tribune, June 19, 1971.

40. Reviewed in, inter alia, Pravda, January 17, 1967; Nedelia, no. 12 (1968), p. 14; Komsomolskaia pravda, April 3, 1968; Okhrana truda i sotsialnoe strakhovanie, no. 11 (1969), p. 32.

41. Radio Moscow-1, 0810 GMT, May 25, 1970.

42. See, for instance, Nash sovremennik, no. 3 (1963) and no. 6 (1963); Literaturnaia gazeta, no. 3 (1965), no. 13 (1965), and no. 21 (1965).

43. Komsomolskaia pravda, April 3, 1968.

44. Sovetskaia Rossiia, May 24, 1970.

45. Molodaia gvardiia, no. 6 (1970), pp. 266-74.

46. Vestnik moskovskogo universiteta: Geografiia, no. 3 (1971), p. 4.

47. New York Times, December 24-25, 1964.

48. Ibid., September 13, 1965.

49. International Herald Tribune, March 14, 1966.

50. Cited in Orbis, Spring 1970, p. 136.

51. New York Times, November 30, 1971.

52. International Herald Tribune, October 8, 1971.

53. UPI, December 14, 1971.

54. See Sovetskaia Rossiia, August 14, 1971.

55. Khimiia i zhizn, no. 1 (1970), p. 55.

56. Business Week, April 10, 1971.

57. Sotsialisticheskaia industriia, August 15, 1970.

58. See, for instance, Literaturnaia gazeta, no. 23 (1970), p. 11.

59. Harvard Bulletin, April 13, 1970.

60. United Nations, Economic Commission on Europe, Economic Survey of Europe in 1971 (provisional version), Part I, p. 182.

61. Nedelia, no. 34 (1967), p. 14.

62. Sovetskaia iustitsiia, no. 2 (1971), p. 1.

63. Literaturnaia gazeta, July 12, 1967.

64. Ekonomicheskie nauki, no. 10 (1971), p. 41.

65. Pravda, October 7, 1968.

66. Nedelia, no. 12 (1968), p. 14.

67. Ekonomicheskaia gazeta, no. 40 (1969), p. 17.

68. Die Welt, January 25, 1972.

69. Komsomolskaia pravda, April 3, 1968.

70. Khimiia i zhizn, no. 1 (1970), p. 55.

71. According to Nedelia, no. 9 (1971), p. 21.

72. Cited in Narodnoe khoziaistvo Kazakhstana, no. 2 (1968), p. 84.

73. Radio Moscow-1, 0900 GMT, November 28, 1971.

74. Literaturnaia gazeta, no. 28 (1967), p. 14.

75. TASS, August 7, 1969.

76. Radio Moscow-1, 0915 GMT, March 23, 1971.

77. Komsomolskaia pravda, April 3, 1968.

78. Cited in New York Times, September 4, 1971.

79. Pravda, April 6, 1971.

80. Sotsialisticheskaia industriia, August 5, 1970.

81. Cited in Posev, no. 6 (1970), p. 37.

82. Gorodskoe khoziaistvo Moskvy, no. 6 (1971), p. 41.

83. Komsomolskaia pravda, April 3, 1968.

84. Ekonomicheskaia zhizn SSSR, Part I, Moscow, 1967, p. 55.

85. Ural, no. 6 (1970), p. 87.

86. See Pravda, December 11, 1970 and Trud, January 9, 1971.

87. I.e., January 1969 and June 1971: see Pravda, October 27, 1971.

88. For automobiles, "clean air" is now defined as a maximum of nine parts of carbon monoxide per million parts of air during an eight-hour period. Hydrocarbons are limited to 24 ppm for a maximum of three hours. For industry, "clean air" now means a maximum of 80 micrograms of sulfur oxide per cubic meter of air and 75 micrograms pcm of particulars as an annual mean. According to the proposed law, the states have until the end of 1972 to present plans for compliance and, if these are accepted by the EPA, until 1975 to carry them out. Automobile manufacturers have until 1985 to reduce the 1970-level emissions by 90 percent.

89. Izvestiia, October 3, 1968 and Pravda, September 24, 1971.

90. See Trud, January 9, 1971.

91. Literaturnaia gazeta, no. 10 (1971), p. 11.

92. Planovoe khoziaistvo, no. 7 (1970), p. 78.

93. See, for instance, Promyshlennost' Belorussii, no. 6 (1971), p. 82.

94. TASS, April 25, 1968.

95. Selskaia zhizn', January 18, 1970.

96. Izvestiia, July 24, 1970.

97. See Sovetskaia Latviia, December 14, 1969.

98. See, for instance, Sovetskaia Rossiia, November 14, 1969, and New York Times, February 13, 1972.

99. Cited in Nedelia, no. 12 (1968), p. 14.

100. Komsomolskaia pravda, April 3, 1968.

101. Ibid.

102. Sovetskaia Latviia, December 14, 1969 (no telephone number was given).

103. Pravda, November 25, 1971.

104. Literaturnaia gazeta, no. 32 (1967), p. 10.

105. Trud, January 9, 1971.

106. Politicheskoe samoobrazovanie, no. 7 (1971), p. 34.
107. TASS, February 1, 1972.
108. Pravda, July 29, 1971.
109. CTK, May 4, 1971.
110. International Herald Tribune, October 16, 1971.
111. The Observer, July 4, 1971.
112. International Herald Tribune, June 19, 1971.
113. Literaturnaia gazeta, no. 48 (1971), p. 24.
114. Literaturnaia gazeta, no. 23 (1970), p. 11.
115. Narodnoe khoziaistvo Kazakhstana, no. 2 (1968), p. 83.
116. Nedelia, no. 9 (1971), p. 21.
117. New York Times, July 17, 1971.
118. Frankfurter Allgemeine Zeitung, August 31, 1966.
119. Neue Zuercher Zeitung, December 18, 1971.
120. Allgemeine Fischwirtschaftszeitung, January 18, 1972.
121. The Times (London), April 29, 1971.
122. TASS, June 1, 1971.
123. Radio Moscow for Seamen, 1230 GMT, October 23, 1971.
124. Nedelia, no. 12 (1968), p. 14.

3

**AIR POLLUTION
IN THE USSR**
Victor L. Mote

More than two centuries have passed since the inception of industrialization, but it has been only in the last half-century that wide
concern has been expressed for the environmental consequences of
rapid industrial progress.[1] For many years Soviet scientists and
ideologists have disparaged capitalism for its "innate" inability to
cope with environmental pollution. In contrast, they have boasted
that such diseconomies could not occur in the Soviet Union because
the contradictions between the public nature of conservation and the
private interests of industry are resolved "only under socialism."[2]
And yet in light of the abundance of news items, journal articles, and
books concerned with environmental problems that have been published
in the Soviet Union during the past decade, it is apparent that pollution
of all sorts may be found in a socialist country as well as in a capitalist
one.

This disparity between the theory and the practice of socialism
has prompted a new wave of criticism from American observers of
Soviet resource and environmental management policies. Highlighting
the parade of papers, articles, and now books, is Marshall Goldman's
The Spoils of Progress, in which it is stated that "the USSR has environmental disruption that is as extensive and severe as ours."[3]
In the January 1972 issue of Harper's Magazine, Peter F. Drucker
averred that "no American city can truly compete in air pollution
with [among others] Moscow."[4] Although there is considerable merit
in much of the Goldman book and the Drucker article, these statements

Acknowledgment is given to Professor W. A. Douglas Jackson
for encouraging this study and to Professor Phillip Bacon for his
helpful emendations.

deserve clarification. Otherwise, it appears that in our frustration with our own environmental difficulties, we are inclined to stretch the truth in the case of our competitors in the USSR. It is certainly not amiss to deflate Soviet illusions about the ability of socialism to maintain a clean and healthy environment in the midst of incredible industrial progress, but to spread in turn an illusion in this country of comparable baselessness is not in the best interests of science nor of knowledge in general.

THE SCOPE OF SOVIET AIR POLLUTION

In his recent relatively dispassionate assessment of Soviet environmental problems, Keith Bush, Senior Economist of Radio Liberty in Munich, reported that air pollution in the USSR has yet to become a major problem.[5] This pronouncement by a Westerner may be augmented by any number of similar statements attributable to Soviet sanitary officials, to whom water pollution is considered a much more serious environmental threat. What, then, is the magnitude of Soviet air pollution?

Not long ago a brief report on air pollution in the USSR was released by the World Health Organization (WHO). The document had been written by Professor K. A. Bushtuyeva, Chief of the Department of Communal Hygiene at the Institute for Advanced Medical Train ing in Moscow. In this report, previously unpublished statistics on atmospheric contamination created by major industrial sectors of the Soviet economy were made available to Western scientists (Tables 3.1 and 3.2).

In the late 1960s most of the air pollution over the Soviet Union was generated by seven branches of industry, including automobile transportation (see Table 3.1). Generation of thermal electric power (P) and ferrous metallurgy (F) accounted for more than half of the total contamination, followed by oil refining and related petroleum industries (O), vehicular transportation, including trucks, cars, and buses (A), nonferrous metallurgy (N), construction materials, chiefly cement production (C), and a small but growing chemical industry.

These industries were responsible for the introduction of five major by-products to the atmosphere (see Table 3.2). Of these five, carbon monoxide (CO), total suspended particulates (TSP) or dust, as they are often referred to in the Soviet Union, and sulfur dioxide (SO_2) composed well over 75 percent of the aerosols. Hydrocarbons (HC), nitric oxides (NO_x), and other pollutants constituted less than one-eighth of the total.

TABLE 3.1

Estimates of Air Pollution Produced by
Various Soviet Industries, 1968-69

Industrial Branch	Percentage of Total
Fuel-burning power stations (P)	27.0
Iron and steel industry (F)	24.3
Oil-producing and petrochemical industry (O)	15.5
Automobile transport (trucks, buses, and cars) (A)	13.1
Nonferrous metallurgy (N)	10.5
Construction materials industry (C)	8.1
Chemical Industry*	1.3

*Statistics not included in the study.

Source: United Nations, World Health Organization, Some Problems in Improving Environmental Sanitation in the USSR with Special Reference to Air Pollution, by K. A. Bushtuyeva (WHO/AP/71.37), 1971, p. 4.

U.S. AND USSR AIR POLLUTION COMPARED

Assuming that air pollution abatement technologies are even remotely similar in the two countries, the gross emission of atmospheric contamination in the Soviet Union should be considerably less than that of the United States. In 1971 the estimated gross national product of the USSR was nearly $600 billion, roughly 60 percent of that computed for the United States.[6] (Termed "national income," the Soviet gross national product does not include the same variables as our own and must be adjusted to suit our conditions.) Still further disparities are revealed once the capacities of the seven major polluting sources in each country are determined and compared. For instance, in 1970, in terms of kilowatt-hours, only 55 percent as much coal-fueled electricity was generated in the Soviet Union as in the United States.[7] In the same year more than 3.5 billion barrels of crude petroleum were produced in the United States, nearly 1 billion barrels more than the 2.6 billion barrels extracted in the USSR.[8] Although the size of the Soviet vehicular stock can only be estimated, a 1970 figure of 6.5 million trucks, buses, and cars, or 27 vehicles

TABLE 3.2

Industrial Air Pollutants in the USSR, 1968-69
(percentages of total air pollution)

Air Pollutants	Percentage of Total
Carbon monoxide (CO)	31.9
Total suspended particulates (TSP)	28.3
Sulfur dioxide (SO_2)	27.0
Hydrocarbons (HC)	10.7
Nitric oxides (NO_x)	1.1
Others	0.5

Source: United Nations, World Health Organization, Some Problems in Improving Environmental Sanitation in the USSR with Special Reference to Air Pollution, by K. A. Bushtuyeva (WHO/AP/71.37), 1971, p. 5.

per thousand persons, seems reasonable. This compared to 105 million cars and trucks alone in the United States, or 514 vehicles per thousand persons.[9] In fact, even with the addition of the new output of the Togliatti plant, by 1975 the Soviet automobile stock may only equal the United States output in 1917.

Turning to nonferrous metallurgy, the USSR in 1970 lagged behind the United States in the smelting of copper by some 1 million tons (U.S.: 1,641,338; USSR: 630,000), in lead by nearly 200,000 tons (U.S.: 666,730; USSR: 485,000), in zinc by more than 200,000 tons (U.S.: 877,811; USSR: 672,000), and in aluminum by close to 3 million tons (U.S.: 4,263,000; USSR: 1,578,000).[10] By 1970, only in cement manufacturing had the production of the United States been overtaken by Soviet industry.[11] In 1971, the USSR also led in ferrous metallurgy. By the end of that year 132.8 million tons of steel had been smelted in the Soviet Union compared to 120 million tons in the United States.[12]

Percentages of the crude weights of air pollution in the United States and the USSR have been obtained and compared (Table 3.3). Assuming the data can be trusted, the volume of SO_2 emitted from Soviet combustion processes during the late 1960s was 16 million tons.[13] Given the overall percentages of pollutants in the atmosphere of the USSR at that time (Table 3.2), crude statistics for the remaining emissions may be calculated (Table 3.4).

The U.S. Council on Environmental Quality has said:

TABLE 3.3

Crude Weights of Air Pollution Emissions
in the U.S. and USSR, late 1960s
(in percentage)

Country	TSP	SO_2	CO	HC	NO_x
U.S.[a]	13	15	47	15	10
USSR[b]	28	27	32	11	1

[a]Computed from U.S. Department of Health, Education, and Welfare, Nationwide Inventory of Air Pollution Emissions 1968 (Raleigh, N.C.: Public Health Service, 1970), p. 3.
[b]See Table 3.2.

TABLE 3.4

Crude Weights of Air Pollution Emissions
in the U.S. and USSR, late 1960s
(10^6 tons/year)

Country	TSP	SO_2	CO	HC	NO_x	Total
U.S.[a]	28	33	100	32	21	214
USSR[b]	17	(16)	19	7	1	60

[a]Actual figures are provided in Table 3.3.
[b]The figure in parentheses was used as the base figure for computing the remainder of the table; see note 13.

41

The weight of air pollution emissions is only a rough measure of air pollution. Indeed, the geographic concentration of pollution sources and the dispersion of the pollutants once they leave the sources determine air quality. Also, weight does not take into account the effects of a pollutant. For example, [this table] considers all particulates as a single category, although the environmental impact of very small particles which add little to total weight, differs markedly from the larger particulates. (It takes 1,000 particles, 0.5 microns in diameter, to equal the weight of 1 particle, 5 microns in diameter, of the same material. Yet one ton of fine particles in the air reduces visibility 25 times as much as one ton of larger particles. And finer particles are also more of a health hazard.)[14]

Although summing the five different pollutants does not enhance our understanding of the relative air quality of the two countries, it does illustrate the comparative volumes of atmospheric contamination. The crude air pollution emissions for the late 1960s in the United States totaled 214 million tons per year compared to 60 million tons in the Soviet Union (roughly 28 percent of its American counterpart).

Recently two CIA employees, David Carey and Robert Dockstader estimated potential volumes of Soviet air and water pollution by means of an input-output model developed by Wassily Leontief.[15] Carey and Dockstader noted:

Assuming the emissions from all mobile sources account for half of the air pollution in the U.S., the output of air pollutants from industrial sources and automobiles in the USSR in 1970 may be . . . one-quarter of the U.S. level.[16]

Interestingly, the results of this study and those of the CIA were derived independently and without knowledge of each other. On the basis of this research, the volumes of Soviet and American air pollution cannot be said to coincide even in the remotest sense.

DISTRIBUTION OF POLLUTANTS

In light of what is known about the major polluting industrial branches in the USSR, insight into the spatial distribution of their associated atmospheric wastes and their physical and chemical composition may be obtained. By analyzing the locations of each given major polluting sector in the Soviet Union, for example "fuel-burning power stations," upwards of 379 potential atmospheric problem areas containing at least one major polluting source may be distinguished. Unfortunately, the state of the air pollution control technology at all of these sites cannot be ascertained at this moment nor can one be certain that this technology is functioning properly. Otherwise, air pollutant indices devised and utilized by the U.S. Public Health Service and the Environmental Protection Agency could be applied analogously to the USSR and some measure of the real pollution output could be determined. Compounding the insufficiency of specific "by-factory" abatement data is the lack of accurate production data for individual sources in certain industrial sectors, such as ferrous and nonferrous metallurgy. Both of these deficiencies render purely speculative any attempt at computing specific volumes of air pollution over Soviet cities and towns.[17]

Major concentrations of polluting centers are located in the Central Industrial District, the Donbass-Dnieper Bend, the Central and Southern Urals, the Samara Bend, the Central Transaucasus, the Tashkent-Fergana Basin, the Kuzbass, and the Angara-Baykal Region. Of course, these regions include vast areas and should not be expected to be polluted everywhere at a constant intensity. On the other hand, under given atmospheric conditions, the air pollution may be detected over sizable tracts of land. (See Map 3.1.)

The principal physical and chemical components of the atmospheric contamination vary according to site and region owing to the disproportional representation of the major polluting industrial branches. Fuel-burning power stations emit chiefly TSP, SO_2, and NO_x. In ferrous metallurgy the basic problems lie with TSP, SO_2, and CO. Petroleum refineries pollute the atmosphere with HC and SO_2. Automobile transportation, one of the oil industry's largest consumers even in the USSR, emits CO, HC, and NO_x. Nonferrous metallurgy is culpable for large quantities of pollution by TSP, SO_2, and fluorides (F). The cement industry is notorious for its emissions of TSP. Considering these factors and knowing the distribution of the given industrial branches, the previously cited regions may be differentiated according to their principal pollutants.

Central Industrial District

Located within this region, the manufacturing hub of the Soviet Union, are some 26 major polluting sites.[18] Reflecting the varied industrial composition of the Center, all five major pollutants including TSP, SO_2, CO, HC, and NO_x should be represented. The relatively high proportions of the latter three pollutants may be accounted for by potentially the greatest regional stock of automobiles.[19] The

MAP 3.1

Air Pollution in the USSR

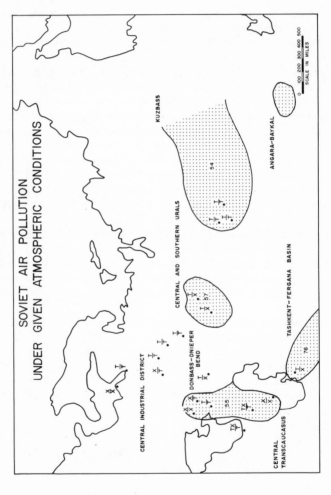

SOVIET AIR POLLUTION
UNDER GIVEN ATMOSPHERIC CONDITIONS

CENTRAL INDUSTRIAL DISTRICT

DONBASS-DNIEPER
BEND

CENTRAL
TRANSCAUCASUS

CENTRAL AND SOUTHERN URALS

KUZBASS

TASHKENT-FERGANA BASIN

ANGARA-BAYKAL

0 100 200 300 400 500
SCALE IN MILES

Source: M. Ye. Berlyand, "The Climatological Aspects of Investigation of Atmospheric Contamination with Atmospheric Wastes," in M. Budyko et al., eds., Modern Problems of Climatology (Wright-Patterson Air Force Base, Ohio: Foreign Technology Division, 1967), p. 310. By permission.

comparatively low emission of SO_2 is a result of the scarcity of the ferrous, nonferrous, and petroleum industries. The most polluted subzone within the region should be found in the vicinity of the Tula-Kosaya Gora ferrous metallurgical complex. Encircling this subregion and considerably enhancing its atmospheric contamination are large power plants burning brown coal situated at Shchekino, Novomoskovsk, Cherepet', and Kashira. Owing to the high ash and sulfur contents of the fuel used in these plants (an average of 28 percent and 3.3 percent, respectively), much of the particulate matter and SO_2 detected in the atmosphere of the Central Industrial district no doubt originates in the region south of Moscow.

Moscow itself has been the scene of a massive air pollution clean-up campaign. In the past fifteen years more than 7,000 purification installations have been set up in the capital city alone. As a consequence of these efforts and the introduction of natural gas and oil heat since 1957, air pollution in the city has been reduced by three to four times.[20] These trends were reflected in the measurements taken as early as ten years ago (Table 3.5). At that time reductions from the 1956 levels for dust and SO_2 averaged 50 percent or more, an obvious improvement despite the fact that most areas of the city still registered pollution levels that were higher than the maximum permissible concentrations for a 24-hour period (dust: 0.15 mg/m^3; SO_2: 0.15 mg/m^3). Ostensibly, these norms currently are being met.

Donbass-Dnieper Bend

Easily comprising more than 10 percent of the country's manufacturing, the Donbass-Dnieper Bend in 1970 produced more than 40 percent of the USSR's pig iron, steel, and rolled metal, and some 75 percent of the nation's steam and electric locomotives.[21] There are no less than 32 major polluting sources in the region, particularly ferrous metallurgy. Owing to this emphasis, air pollution from TSP, SO_2, and CO is unquestionably severe, particularly in the vicinities of Dnepropetrovsk, Krivoy Rog, and Zaporozh'ye from the Dnieper River Valley west and east of the river in and around the cities of Donetsk, Makeyevka, and Yenakiyevo. Some pollution by fluorides also may be anticipated in the presence of the Zaporozh'ye aluminum refinery whose estimated capacity in 1969 was 80,000 tons.[22]

During the early 1960s atmospheric pollution in the Donbass was curtailed substantially by the introduction of gas-fueled household boilers and purifying installations. For instance, in Voroshilovgrad (then Lugansk) SO_2 concentrations were reduced by a factor of 4 and TSP by a factor of 24. In Donetsk, SO_2 and TSP pollution were cut in half. TSP losses at the Kramatorsk cement works also were

TABLE 3.5

Dust and Sulfur Dioxide Concentrations in the Ambient Air of Moscow, 1956-62
(in mg/m^3)

| Area | Season[a] | | | | | | | |
| | 1956 | | 1958[b] | | 1960 | | 1962 | |
	1	2	1	2	1	2	1	2
Dust Concentrations								
Industrial	0.98	1.0	0.36	0.46	0.25	0.45	0.22	0.30
Residential	0.74	0.67	0.25	0.28	0.25	0.32	0.20	0.36
Highway	0.77	0.97	0.64	0.71	0.11	0.43	0.28	0.51
Railway	0.60	0.96	0.22	0.32	0.39	0.54	0.18	0.18
Parks	0.59	0.66	0.17	0.22	0.10	0.11	0.18	0.18
Sulfur Dioxide Concentrations								
Industrial	0.84	0.82	0.37	0.24	0.31	0.19	0.40	0.25
Residential	0.84	0.51	0.28	0.13	0.23	0.14	0.25	0.14
Highway	—	—	0.52	0.13	0.51	0.30	0.29	0.20
Railway	0.78	0.68	0.56	0.22	0.48	0.23	0.31	0.18
Parks	0.78	0.70	0.60	0.20	0.16	0.06	0.24	0.17

[a]1 = heating season; 2 = nonheating season.

[b]Introduction of gas and oil in district heating and power plants.

Source: M. S. Sokolovskiy, "The Moscow City Sanitary-Epidemiological Station," in United Nations, World Health Organization, Air Pollution Control (WHO/EP/68.2) 1968 p. 32. By permission.

halved after the construction of electrostatic precipitators. The same emissions at another cement plant in Dneprodzerzhinsk were reduced from 100 tons to roughly 14 tons daily.[23] Some evidence would appear to imply that a 35.8 percent reduction in Ukrainian industrial occupational morbidity between 1960 and 1967 evolved at least in part as a consequence of air pollution abatement in the Donbass-Dnieper Bend.[24]

Central and Southern Urals

Possessing nearly as high a percentage of Soviet manufacturing as the Donbass-Dnieper Bend area, the large region herein designated as the Central and Southern Urals accounts for some 45 major polluting sources. The majority of these sources are associated with ferrous and nonferrous metallurgy. Thus, air pollution from TSP and SO_2 is a considerable problem, especially in the cities of Magnitogorsk, Chelyabinsk, Nizhniy Tagil, and Beloretsk.

By far the most polluted city in the region, if not in the Soviet Union, Magnitogorsk receives only one-third as much solar energy (in calories per square meter) as the sanitorium of Yakte-Kule, some 35 miles to the northwest.[25] Contributing to the city's pollution dome are hundreds of tons of TSP, SO_2, and CO emitted from dozens of coke ovens, blast furnaces, and open hearth furnaces smelting more than 12 million tons of crude steel annually.[26] Moreover, if pollution abatement is held constant, these tonnages of by-product should increase during the current Five-Year Plan. By 1975 "Magnitka" is scheduled to produce as many as 7 million more tons of pig iron and 10 million more tons of steel.[27]

Interestingly, during the 1950s and 1960s thousands of Magnitogorsk residents were resettled on the west bank of the Ural River, ostensibly owing to the "appalling housing conditions" in the old section of town on the east bank.[28] Though the role of air pollution in the decision to relocate the populace is difficult to substantiate, the old town manifestly stands on the leeward side of the iron and steel complex, whereas the new town is comfortably situated obliquely out of the path of the prevailing wind direction.

Some effective air pollution abatement measures have been implemented in the Urals, although, it seems, much more slowly than in the Center and the Donbass-Dnieper Bend. Chelyabinsk, for instance, has witnessed the introduction of central heating (eliminating at least 300 domestic furnaces), new air pollution control technology at its numerous plants and factories, and a switch from coal to gas fuel.[29] A new city plan for Chelyabinsk calls for the complete elimination of smoke plumes, the placement of modern, "depolluted" factories within residential zones, and its general transformation from a factory town into a park city.[30] Of course, these are long-term goals.

Nizhniy Tagil is another major city in the Urals in which the installation of air pollution abatement equipment is proceeding. During the 1960s electrofilters were mounted on the rotary kilns and coal-drying drums of the local cement mill, and dust discharges were reduced from 600 tons to 90 tons per day. Emissions of fly ash from the power plant of the rolling stock mill were also curtailed 8 to 1 by means of wet scrubbers. However, TSP and SO_2 pollution from the city's metallurgical combine more than neutralized these advances, thus maintaining the concentrations of these pollutants in the ambient air above the permissible limits.[31]

A major breakthrough in air pollution abatement technology is said to be in operation in the Urals at the Asbest asbestos complex, where as much as 100 tons of deleterious asbestos dust are retained hourly by the pollution control system. This system is so efficient that smokestacks have been obviated in at least one of the plant's facilities, a trend ardently supported by Soviet academicians.[32]

Samara Bend

No doubt the least polluted of the regions mentioned so far and accounting for considerably less than 5 percent of the total manufacturing of the USSR, the Samara Bend is dominated by several large cement plants at Zhigulevsk, Sengiley, Ul'yanovsk, and Novoul'yanovsk, and by the oil and petrochemical industry in Kuybyshev and vicinity. Thus, the most serious atmospheric pollution problems should lie with TSP and the frequently noxious odors of hydrogen sulfide and other refinery by-products.

Central Transcaucasus

This region likewise is less polluted than any of the first three. The bulk of its contamination arises from the emissions of the cement industry at Ararat, Rustavi, Tauz, and Kaspi, a major copper smelting and refining center at Alaverdi, the Rustavi iron and steel foundry, a relatively small aluminum refinery at Yerevan, and the automobiles of Yerevan and Tbilisi. TSP and SO_2 should be the principal pollutants.

Owing to the steep slopes of the Transcaucasus, most of these cities are subject to the phenomenon of air drainage and, as a consequence, surface inversions. Tbilisi suffers from the presence of the Los Angeles type of subsidence inversion for at least half the year, and at times the Georgian capital has been plagued with photochemical smog.[33] To circumvent the inversion problem at the Razdan Regional Electric Station in Armenia, a smokestack 540 feet high has been constructed.[34] Dealing less successfully with the inversion problem, officials at Alaverdi noted in the late 1960s that within a 20 km radius of the copper complex, concentrations of SO_2 and sulfuric acid were two to four times greater than the maximum permissible concentrations.[35]

Tashkent-Fergana Basin

Although this region doubtless possesses less than 3 percent of the total manufacturing of the Soviet Union, it comprises two well-known nonferrous metallurgical centers at Chimkent and Almalyk, a minor ferrous metallurgical operation at Bekabad, sizable cement plants at Chimkent, Khilkovo, Angren, Sastobe, Akhangaran, and Kuvasay, and two relatively small petroleum refineries at Fergana and Vannovskiy. Consequently, the region is characterized by much particulate pollution and isolated instances of contamination from SO_2 and HC.

The Tashkent-Fergana region also is plagued with periodic outbreaks of natural dust pollution. It is known to have some of the highest turbidity indices in the Soviet Union because of the combination of natural and man-made aerosols. A study in 1965 revealed that the highest of these indices in Soviet Central Asia were recorded in the vicinities of Tashkent, Andizhan, and Fergana.[36] Indeed, the surprisingly extensive zone of pollution over Central Asia owes its origins more to natural than to man-made causes (see Map 3.1).

Kuzbass

In terms of intensity the air pollution of the Kuznetsk Basin may rank below only that of the Donbass-Dnieper Bend and Ural regions. The Kuzbass is an interesting contrast to the widespread natural pollution of Central Asia, for most of the atmospheric contamination in this area arises from human activities: the smelting of iron and steel and the refining of aluminum at Novokuznetsk; the coal-fed power production from plants near Osinniki, Belovo, Kemerovo, Mezhdurechensk, and Novokuznetsk; the reduction of zinc at Belovo; and coke-chemical production at Kemerovo. All of these industries contribute enormous quantities of TSP, SO_2, CO, and F to the atmosphere. In fact, air pollution is so common that the presence of "yellow smoke" in the air over the industrial centers is taken for granted by many of the region's residents.[37]

Angara-Baykal Region

That this rather remote area of the USSR would have any significant air contamination may come as a surprise to some readers.
However, the statement of one qualified Soviet observer leaves no
room for doubt: "Since the construction of the Irkutsk Dam, the Angara,
remaining unfrozen in the vicinity of the dam for the entire winter,
enables the condensation of vapors which coupled with the smoke from
the factories form deadly fogs reminiscent of London."[38] The factories
alluded to in the statement comprise the huge Shelikhov aluminum
refinery; the Angarsk petroleum, petrochemical, and cement operations;
and a medium-sized coal- and oil-burning thermal electric facility at
Irkutsk. Thus, TSP, SO_2, and abundant quantities of F are the primary
pollutants found in the atmosphere of this region.

CONCLUSIONS

Though it may be a common problem, even an extraordinarily
serious one in such metallurgical centers as Magnitogorsk, Krivoy
Rog, and Novokuznetsk, air pollution in the Soviet Union is certainly
not as extensive and severe as it is in the United States. This is not
a consequence of better pollution control technology nor of better
management; rather, it is the result of a still lagging industrial economy and the relative absence of the automobile. Given another decade
or two these factors may be compromised, but at the moment Soviet
air pollution represents less than one-third the volume of its American
counterpart.

Air pollution in the USSR would appear to be problematic in at
least 379 sites, each reflecting one or more major polluting sources.
Based on the spatial arrangements of these sources, some eight air
pollution regions have been determined: the Central Industrial District; the Donbass-Dnieper Bend; the Central and Southern Urals; the
Samara Bend; the Central Transcaucasus; the Tashkent-Fergana
Basin; the Kuzbass; and the Angara-Baykal Region. In the absence
of actual production data and known pollution control technology for
each source, speculation concerning which of these regions suffers
most from the problem is a futile exercise. However, with the understanding that power production and ferrous metallurgy contribute
more than half of the air pollution in the Soviet Union, one can safely
conclude that either the Urals or the Donbass-Dnieper Bend is the
most contaminated region.

NOTES

1. See R. Murphey, The Scope of Geography (Chicago: Rand McNally and Company, 1966), p. 148; S. W. Allen and J. W. Leonard, Conserving Natural Resources (New York: McGraw-Hill, 1966), p. 147. Although the exact date varies with the given country, the advent of the Industrial Revolution occurred circa 1750. Despite the acknowledgments of environmental degradation as early as the thirteenth century, serious concern for its abatement did not arise until sixty-three persons perished in the wake of a severe air pollution condition near Liege, Belgium, in 1930.

2. I. P. Laptev, Okhrana prirody (Tomsk, 1964), p. 9; M. S. Goldberg, Sanitarnaya okhrana atmosfernogo vozdukha naselĕnnykh mest (Moscow, 1960), p. 4; V. A. Ryazanov, ed., Limits of Allowable Concentrations of Atmospheric Pollutants, Book 2, trans. B. S. Levine (3 vols; Washington, D.C.: U.S. Public Health Service, 1955), p. 144; I. Petryanov, "Quo Vadis?" Soviet Life, November 1970, p. 52.

3. M. I. Goldman, The Spoils of Progress: Environmental Pollution in the Soviet Union (Cambridge, Mass.: MIT Press, 1972), p. 2.

4. P. F. Drucker, "Saving the Crusade," Harper's Magazine, January 1970, p. 70.

5. K. Bush, "Environmental Problems in the USSR," Problems of Communism 21 (July-August 1972), pp. 21-31.

6. See A. Bergson, The Real National Income of Soviet Russia Since 1928 (Cambridge, Mass.: Harvard University Press, 1961). The $600-billion figure was reported in "The Soviet Union: The Risks of Reform," Time, March 29, 1971, p. 30. A near-trillion-dollar figure for the United States appears in L. H. Long, ed., The World Almanac (New York: Newspaper Enterprise Association, 1972), p. 71.

7. As reported in Narodnoe khozyaystvo SSSR v 1970 g. (Moscow, 1971), p. 179, in 1970 Soviet thermal electric plants produced more than 600 billion kilowatt-hours of electricity. Based upon a thorough analysis of sources concerned at least in part with the distribution and composition of Soviet steam electric stations (including TESs and GRESs), a figure of 70 percent was selected to represent the volume of electricity generated by coal-fired plants: see A. T. Khrushchev, Geografiya promyshlennosti SSSR (Moscow, 1969), p. 248a; I. T. Novikov, Energeticheskiye resursy SSSR: Toplivo-energeticheskiye resursy (Moscow, 1968), p. 38a; T. Shabad, Basic Industrial Resources of the USSR (New York: Columbia University Press, 1969), pp. 24, 95, 97, 104, 106, 108, and scattered references; V. A. Shelest, Ekonomika razmeshcheniya elektroenergetiki SSSR (Moscow, 1965), pp. 199-237; A. N. Voznesenskiy et al., Atlas razvitiya khozyaystva i kul'-tury SSSR (Moscow, 1967), pp. 18-19.

In the USSR roughly 420 billion kilowatt-hours of electricity were coal-fueled in 1970; in the United States in 1970 approximately 751 billion kilowatt-hours (46 percent of 1.64 trillion KWH) were created by coal-fired plants. See U.S., Department of Commerce, Statistical Abstract of the United States 1971 (Washington, D.C.: U.S. Government Printing Office, 1971), p. 497.

Sulfur and ash contents of the coal utilized in the two countries are similar: the average sulfur content is 2.5 percent and the average ash content is 10 to 20 percent. See U.S., Department of Health, Education, and Welfare, Emissions from Coal-Fired Power Plants by S. T. Cuffe and R. W. Gerstle, Public Health Service Publication No. 999-AP-35 (Washington, D.C.: U.S. Government Printing Office, 1969), pp. 12, 16; and O. de Lorenzo, Combustion Engineering (New York: Combustion Engineering-Superheating, Inc., 1950), pp. A7-A10.

8. U.S., Department of the Interior, Minerals Yearbook 1970 (Washington, D.C.: U.S. Government Printing Office, 1972), p. 923.

9. D. Carey and R. Dockstader, "USSR: Economic Implications of Environmental Disruption" (paper presented at the Fifth National Convention of the American Association for Advancement of Slavic Studies, Dallas, Texas, March 16, 1972), p. 7.

10. U.S., Department of the Interior, op. cit., pp. 499 (note 8), 665, 1204, 179, 181.

11. U.S., Department of the Interior, op. cit., pp. 292-293; (note 10); tonnages were 558,192,000 barrels for the Soviet Union and 406,721,000 for the United States.

12. G. W. Graham, "For First Time, Russ Pour More Steel Than U.S.," Houston Chronicle, January 27, 1972, section 1, p. 18.

13. I. V. Komar, "Obmen veshchestv 'priroda-obshchestvo-priroda' i nekotoryye voprosy yego optimizatsii," Izvestiya Akademii Nauk SSSR, seriya geograficheskaya, no. 5 (1969), p. 48. Roughly 7 million tons were generated from electric power production in 1967. This latter figure was found in a special print-out requested from the Air Pollution Technical Information Center, Durham, North Carolina, October 1, 1970, which cited and abstracted "Chistyy Vozdukh," NTO-SSSR 10, no. 12 (1968), pp. 17-18.

14. U.S., Executive Office of the President, Environmental Quality: The Third Annual Report of the Council on Environmental Quality (Washington, D.C.: U.S. Government Printing Office, 1972), p. 5.

15. Carey and Dockstader, op. cit., p. 24n; W. Leontief, "Environmental Repercussions and the Economic Structure: An Input-Output Approach," Review of Economics and Statistics, August 1970, pp. 262-71.

16. Carey and Dockstader, op. cit., p. 17n.

17. Though fraught with assumptions and estimations, the attempt has been made by V. L. Mote in "The Geography of Air Pollution in the USSR" (Ph.D. dissertation, University of Washington, 1971).

18. See R. E. Lonsdale and J. H. Thompson, "A Map of the USSR's Manufacturing," Economic Geography 34 (January 1960), pp. 36-52. In 1960 the Center, excluding Gor'kiy, reflected about 17.3 percent of all Soviet manufacturing. Today this region no doubt continues to account for more than 15 percent of the manufacturing of the country.

19. This assumption is based on remarks in M. Seeger, "Wheels for Ivan," The Washington Post, June 18, 1972, B3. According to Seeger, Moscow possesses the overwhelming number of cars, perhaps 360,000 (both private and official), as opposed to Leningrad's 85,000 and Kiev's 19,000. In a private letter to me on July 11, 1972, Seeger said his data were derived from Literaturnaya gazeta, September 22, 1971.

20. This information was announced at the September 1972 session of the Supreme Soviet. "Berech' i umnozhat' bogatstv strany," Pravda, September 21, 1972, p. 2.

21. Narodnoye khozyaystvo, op. cit., p. 70. This is one of the country's smallest aluminum refineries.

22. U.S., Department of the Interior, Minerals Yearbook 1969 (Washington, D.C.: U.S. Government Printing Office, 1971), p. 167.

23. M. V. Kryzhanovskaya et al., "Opyt ozdorovleniya atmosfernogo vozdukha v gorodakh i selakh Ukrainskoy SSR," Gigiyena i sanitariya 33 (May 1968), p. 105.

24. A. V. Pavlov, "Sanitarnyye meropriyatiya: na sluzhbu okhrany zdorov'ya naseleniya," Gigiyena i sanitariya 33 (November 1968), p. 67.

25. B. V. Rikhter, "Natural Ultraviolet Radiation under Different Conditions of Atmospheric Air Pollution," in USSR Literature on Air Pollution and Related Occupational Diseases, Vol. 8, trans. B. S. Levine (10 vols.; Washington, D.C.: U.S. Public Health Service, 1963), p. 201.

26. T. Shabad, "News Notes," Soviet Geography: Review and Translation 11 (December 1970), p. 877.

27. "Magnitka: pyatiletke," Pravda, January 4, 1972, p. 1.

28. M. G. Clark, "Magnitogorsk: A Soviet Iron and Steel Plant in the Southern Urals," in Focus on Geographic Activity: A Collection of Original Studies, R. S. Thoman and D. J. Patton, eds. (New York: McGraw-Hill, 1964), p. 129.

29. V. Varavka, "Seryye oblaka," Ekonomicheskaya gazeta, no. 4 (January 1968), p. 40.

30. Yu. Shlakov, "Samotsvet i yego oprava," Pravda, November 29, 1972, p. 6.

31. V. B. Oparin, "Purification of Atmospheric Air of Contaminants from Industrial Discharges," in AICE Survey of USSR Air Pollution Literature, Vol. 2, M. Y. Nuttonson, ed. (15 vols.; Silver Spring, Md.: American Institute of Crop Ecology, 1969), p. 3.

32. M. Vasin, "Truba podayet v otstavku," Pravda, February 24, 1971, p. 2; and I. Petryanov, op. cit., pp. 50-53.

33. A. Bukhnikashvili and A. Balabuyev, "Za chistyy vozdukh v Tbilisi," Zarya vostoka, May 23, 1968, p. 4.

34. G. Arakelyan, "Razdan darit svet," Pravda, March 16, 1971, p. 1.

35. T. G. Torosyan et al., "O zagryaznenii atmosfernogo vozdukha g. Alaverdi," abstracted in Referativnyy zhurnal: geografiya. L71 okhrana prirody i vosproizvodstvo prirodnykh resursov, no. 2 (February 1971), p. 27. The maximum permissible concentration for sulfuric acid is 0.3 mg/m^3 at any one time and 0.1 mg/m^3 over a 24-hour period. In terms of the philosophy behind them, maximum permissible concentrations differ considerably from standards. See Mote, op. cit., pp. 77-79 (note 16).

36. M. V. Sitnikova, "Atmospheric Turbidity in Central Asia," Trudy sredneaziatskogo nauchno-issledovatel'skogo gidrometeorologicheskogo instituta, no. 22 (1965), pp. 51-59 (in translation on JPRS microprint #1004-1966).

37. V. Chivilikhin, "Kak vam dyshitsya, gorozhane?" Literaturnaya gazeta, August 9, 1967, p. 10.

38. D. L. Armand, Nam i vnukam (Moscow, 1964), p. 132.

4

SOIL EROSION
IN THE CENTRAL RUSSIAN
BLACK EARTH REGION
Ihor Stebelsky

The Central Russian Black Earth Region, located to the northeast of the Ukrainian SSR and south of Moscow (see Map 4.1) has an agricultural potential, in the Russian Republic, surpassed only by the North Caucasus. Yet this region, though known for its rich chernozems,* has some of the most severely eroded lands in the Russian Republic that reduce agricultural land productivity and thwart Soviet attempts to intensify agriculture in the region. In the Soviet literature the cause of this damage is related, first and foremost, to the backward agricultural practices of Russian peasants before collectivization. However, little is said of the soil and its fate in the postcollectivization period, when the soil, despite some widely publicized conservation programs, continues to suffer from erosion.

The purpose of this chapter is to describe the environmental setting and the extent of soil erosion, to evaluate the causes of environmental deterioration, and to assess the impact of socialized agriculture on the environmental quality of the region.

SETTING

The study area, centrally located in the East European Plain, is bisected by the southward-flowing Don River (see Map 4.2). Its western half is the rolling Central Russian Upland, with a strongly entrenched river system and a prevailing elevation of 200 to 250 meters

---

*Chernozems refers to black soil or black earth. The chernozem region is assumed to be, perhaps, the richest, most fertile land in the USSR.

55

MAP 4.1

Central Russian Black Earth Region

Source: Composed by the author.

above sea level. East of the Don, except for the Don-Voronezh side and the Kalach Upland, stretches the flat to gently undulating Tambov Lowland, with elevations from 125 to 175 meters above sea level.[1]

The topography and hydrography of the region developed on very slightly folded and occasionally faulted permeable sedimentary rocks: limestones, chalks, and sandstones. During the Pleistocene only its northern periphery and the Tambov Lowland were glaciated, acquiring a layer of ground moraine that is not always permeable to water. However, most of the surface, particularly in the nonglaciated upland, obtained a deep cover of loess.* This material, though permeable and easily erodable, formed an excellent base for the development of fertile gray forest loams and chernozems.

The rich soils supported lush natural vegetation. In the northern and western peripheries were thick broadleaf forests, thinning southward into a parkland of oak groves and meadows and open feathergrass steppe in the southeastern extremity. Such zonation reflected the availability of moisture, inasmuch as rainfall diminished from 600 mm (about 24 inches) in the northwest to less than 450 mm (about 17 inches) in the southeast,[2] while summer temperatures increased from 18°C (65°F) along the Oka River to 22°C (72°F) in the southeastern extremity of the region.[3] However, the Central Russian Black Earth Region was also susceptible to drought (the Russian sukhovei) and to torrential local rainstorms in the summertime, which were particularly intense in the Central Russian Upland. Once the natural vegetation was cleared, the soil became directly exposed to running water, dessication, and dry wind.

The problem of soil erosion in the Central Russian Black Earth Region was recognized by a number of Russian geographers and soil scientists as early as the nineteenth century.[4] However, a comprehensive evaluation of the extent of the damage was not completed until the mid-twentieth century. In 1945 Guzhevaia produced the first map of gullies in the Central Russian Upland. She found that 65 percent of that land area was covered either by a medium (0.2 to 0.4 km of gully per km^2) or a dense (0.5 to 1.2 km of gully per km^2) gully network. Because the greatest densities occurred along the high right bank of the Don River, in the Don-Voronezh interfluve, and throughout the Kalach Upland, Guzhevaia concluded that this pattern corresponded to the distribution of sharp local relief, in the order of 100 meters or greater.[5]

Subsequent efforts of Soviet geographers yielded maps showing both the types and the degrees of soil erosion.[6] Although the detail

---

*Loess is a soft, homogeneous, porous type of soil, having grains smaller than sand and considered to be of wind-blown origin.

MAP 4.2

A Detailed Map of the
Central Russian Black Earth Region

Source: Composed by the author.

shown on the two maps was classified in different ways, the patterns were consistent with each other and resembled the one described by Guzhevaia. The highest densities of gullies (1.0 to 3.0 km per km^2) and sloping lands with moderate to severe soil erosion occurred along the right bank of the Don River, near Lipetsk, and on sloping lands of the Kalach Upland. In all these areas there was practically no level upland not affected by erosion.

Intermediate gully density (0.5 to 1.0 km per km^2) on sloping land with moderate to severe soil erosion covered nearly half of the Central Russian Upland. The remaining parts of the upland had a low gully density (up to 0.5 km per km^2) though the level of soil erosion was moderate.

The least soil erosion was encountered in the Tambov Lowland, where nearly half of the land is level and little affected by the denudation processes.

In general, more than half of the area of the Central Russian Upland had soil either moderately or severely eroded.[7] In other words, in the case of moderate soil erosion most of the A horizon had been washed away and the remaining B horizon was cultivated; in the case of severe erosion, the upper layers of the B horizon were also lost, and only the remnants of the B horizon and the parent material itself were cultivated.[8]

CAUSES OF SOIL EROSION

The physical factors relating to soil erosion are easily recognized: sharp local relief, steep slopes, long, continuous slopes, easily erodable parent material, clayey soil, high intensity of rainfall, freeze-thaw effects, dessication, and others.[9] All these are operative in the Central Russian Black Earth Region and manifest themselves in the spatial variation of the magnitude of soil erosion described above.

While events relating to natural geomorphic processes can trigger or renew soil erosion, the major and obvious causes of soil erosion in the Central Russian Black Earth Region are anthropogenic.[10] Among them the most serious are the removal of natural vegetation and the cultivation of the land.

In the past, the lush vegetation of the region protected the soil from erosion. The trees of the forest shaded the ground in the spring, allowing for a gradual snow melt, while the ground litter and undergrowth allowed for such a high percolation capacity that surface runoff was negligible. Even the prairies, with their tall stand of grasses, encouraged an even dispersion of snow in the winter and slow melting with little runoff in the spring. Any damage from the meltwaters was further reduced by a shallow but continuous carpet of moss and a dense layer of grass roots in the sod.[11]

The Central Russian Black Earth Region was the first forest-steppe environment to be settled by Muscovites from the north. Consequently, the land was exposed to farming and soil erosion for more than 350 years. In the early stages of settlement, where considerable land was already cleared in the northern periphery, shallows were observed developing in the Don and Voronezh Rivers.[12] In the course of the seventeenth and eighteenth centuries forests were cut, the virgin steppe was burned, and nearly half of the land area was cultivated.[13] By the beginning of the nineteenth century, when most of the land had been cleared and two-thirds of all land cultivated, widespread soil erosion and gullying appeared—a phenomenon that already generated concern in the 1850s. Even so, the degree of soil erosion and the density of gullies continued to grow, spurred on by the encroachment of the plow onto topographically marginal lands.

This exceptional pressure on the land that the region experienced stemmed from continuous population growth and a lack of an appropriate adjustment in resource use. While the population more than doubled in the nineteenth century,[14] there was no commensurate growth in industrial employment and no improvement in agricultural techniques. The only method the peasants knew of increasing their food supply was to extend their grain acreage. In so doing they overtaxed their three-field system of farming, a system that depended on adequate natural pastures for livestock feed and hence adequate manure for the grain fields.

By the end of the nineteenth century almost three-quarters of the land was cultivated, mainly for grains.[15] So great was the need for agricultural land that pastures on river terraces and slopes were plowed while the flood meadows below the villages were frequently cultivated as kitchen gardens. Forests were severely reduced, causing shortages of wood for housing and fuel. The peasants then turned to straw for their domestic needs, diverting it from its chief use as feed. Meanwhile, the continued expansion of grains reduced meadows and hayfields, making this region suffer the greatest decline in livestock,[16] and the lowest standard of living in European Russia.[17]

Sociopolitical factors were partly responsible for land hunger. On the one hand, rural population densities continued to grow as the government discouraged emigration from the region until the last decade of the nineteenth century.[18] On the other hand, the land reform associated with the emancipation (1861) deprived the former serfs of one-tenth to one-fifth of their best tilled land and assigned it to the newly delineated estates.[19] Thus the estate owners could benefit from surplus labor and depressed wages in the region.

The traditional Russian farming techniques further aggravated soil erosion. Shallow plowing by means of the wooden sokha wore out the surface soil structure. The traditional emphasis on grain culture in the three-field system and a shortage of pastures coupled with the distant scattered strips of arable land contributed to inadequate manuring and led to soil depletion. Finally, the Russian peasant communities (obshchiny), in their concern for equity that would ensure each household to get each soil type, laid out their strips of tilled land down slopes—a practice that contributed to severe gully erosion.[20]

SOVIET EXPERIENCE

After the Bolshevik Revolution, according to Soviet writers, conditions became more favorable to combat soil erosion.[21] The land was nationalized, estates were eliminated, and all the small scattered strips of plowland were merged into large collective fields. Mechanization and the use of tractors allowed for deeper plowing, which made fuller use of the soil and increased the percolation ratio.

In the Central Russian Black Earth Region, rural population densities were sharply reduced and land hunger alleviated. New crops and a multifield crop rotation pattern replaced the traditional three-field system, and clean fallow was virtually eliminated.[22]

Perhaps the most spectacular attempt was Stalin's plan for the transformation of nature announced in October 1948, which involved a widespread planting of shelterbelts, the introduction of grass-field crop rotations, and the construction of many ponds in the steppe and forest-steppe regions of European USSR.[23]

Given the political climate of the day, Soviet scientists confidently asserted victory over soil erosion. In 1949, Dmitriev emphasized that agricultural decline suffered before the Revolution was reversed by progress made under the Soviet rule,[24] while Sil'vestrov claimed that the erosion presently evident in the USSR was directly caused by poor agricultural management before the Revolution.[25]

Since Stalin's death, however, the views of Soviet scientists regarding soil erosion have become more accommodating. In 1963, Sil'vestrov observed: "Historically, extensive erosion and deflation in the USSR reflected the faulty use of agricultural lands prior to the October Revolution, but such use has not been completely eliminated even today."[26] Other authors have expressed similar views.[27] Perhaps a more open political environment and new empirical evidence gathered by the scientists contributed to this change.

The disillusionment came several years after the Academy of Sciences of the USSR dispatched an expedition, in 1951, to study the progress made against soil erosion in the regions encompassed by Stalin's shelterbelt afforestation program. The findings of this expedition were published in 1954, at a time when Stalin's successors, in maneuvering for political advantage, opened agricultural policy to criticism.[28]

In his informative article, Armand noted a number of practices that annulled the effectiveness of shelterbelts and had harmful effects on the soil. To begin with, many shelterbelts on the collective farms were neither properly laid out nor adequately managed, thus promoting rather than curtailing gully erosion. Even tree plantings along the edge of the existing gullies, conceived with the intent of arresting gully erosion while maximizing the flatter land above for plowing did more harm than good. The plantings not only failed to absorb the run-off from the fields above but also accumulated snow in the winter for an even greater runoff in the spring, thus rejuvinating gully erosion.[29]

In the meantime, pastures, instead of being utilized along the more eroded edges of the fields, were often concentrated in one massive plot, and their value for soil conservation was wasted.[30]

Furthermore, the collective farm management, in its preoccupation to fulfill the ambitious government plans for sown acreage, overextended its plowland beyond safe limits and onto slopes generally considered too steep for plowing. This neglect for slope and the eroded condition of the soil on it stemmed from a blind observance of outdated local land surveys and religious adherence to the recommendation that fields of no less than 100 hectares were to be delineated for the sake of economic plowing. As a result, the fields were disjointed by ravines or crossed and included all kinds of slopes, each of which would have warranted a different program of management. Fields were not laid out according to contours, and with only huge, unmaneuvrable tractors available for field work, differences in topography were ignored. Indeed, cases were reported where fields were cultivated downslope in order that the tractor brigades save money on gas![31]

Despite the revelation of such glaring shortcomings and the publication of new and more comprehensive instructions, such malpractices, as shown in pictures taken by the author in 1968, continue to this day. Techniques developed for the flat and lightly eroded Tambov Lowland are applied without modification by the collective farms of the Central Russian Upland, increasing the possibility of soil erosion.[32] Because of the Soviet preoccupation with large-scale mechanized farming, the more labor-intensive contour strip cropping, recommended by Voeikov before the Revolution,[33] is dismissed by Soviet experts as not applicable.[34] In the meantime, large fields of plowland generate exceedingly long lines of drainage contributing to a higher volume and velocity of runoff and hence a more intense level of erosion.[35]

Moreover, despite the elimination of clean fallow, cultivated farmland continues to be exposed to sheet and rill erosion. With Khrushchev's corn program and attempts to intensify agriculture, more of the cultivated land is now under row crops, especially corn.

Corn is demanding on soil yet provides no ground cover in the spring. As a result, soil becomes rapidly exhausted and its surface eroded.

In view of such continuing pressures on the land, the party and government have urgently called, in 1967 and again in 1971, to implement measures to prevent soil erosion.[36] Such decrees indicate that erosion remains a serious problem. At the same time, the Soviet approach to combating soil erosion continues to be based on directives from above rather than on a closer relationship with scientific organizations at a local level.

CONCLUSION

The origin of soil erosion in the Central Russian Black Earth Region can be traced to human occupancy and the removal of natural vegetation. Severe soil erosion reached its maximum extent toward the end of the nineteenth century, when the expansion of agricultural production was carried out by expanding cultivation onto marginal lands rather than by improving the outmoded techniques of farming. While the Soviets reduced land hunger and eliminated the traditional peasant ways of farming, their goals in agriculture continue to stress the expansion of sown acreage, and the cultivation of marginal lands continues to this day. In the meantime, big, unwieldy tractors and large-scale, stereotyped operations make it difficult, if not impossible, to introduce contour strip cropping. Although today the Soviet scientists do not place all the blame for present soil erosion on the pre-Revolutionary agriculture, they still claim that the worst damage to the soil was committed under the peasant farming system. This statement, however, cannot be proven, inasmuch as measurements of erosion were not made before the 1950s. Indeed, there is room for speculation that the levels of erosion experienced in the Central Russian Black Earth Region may be just as severe at the present time as they were before the Revolution.

NOTES

1. V. P. Semenov, Rossiia, Vol. 2, Srednerusskaia Chernozemnaia Oblast' (St. Petersburg, 1902), p. 2.

2. A. M. Grin, Dinamika vodnogo balansa Tsentral'no-Chernozemnogo raiona (Moscow, 1965), p. 19.

3. Atlas SSSR (Moscow, 1962), pp. 79 and 82; A. A. Borisov, Klimaty SSSR (Moscow, 1959), pp. 77 and 188-89.

4. V. A. Kipriianov, Zametki o rasprostranenii ovragov v Iuzhnoi Rossii, Zhurnal Glavnogo upravleniia putei soobshcheniia,

Vol. 26, Bk. 4 (1857); P. A. Kostychev, Pochvy Chernozemnoi oblasti Rossii. Ikh proiskhozhdenie, sostav i svoistva (St. Petersburg, 1886); V. V. Dokuchaev, Nashi stepi prezhde i teper' (St. Petersburg, 1892), reprinted in part in V. S. Dmitriev, ed., O travopol'noi sisteme zemledeliia (Moscow, 1949) pp. 30-45; A. A. Izmail'skii, Kak vysokhla nasha step' (Poltava, 1893); A. I. Voeikov, Vozdeistvie cheloveka na prirodu, Zemlevedenie, Books 2 and 4, (1894), reprinted in A. I. Voeikov, Vozdeistvie cheloveka na prirodu, E. M. Murzaev, ed., (Moscow, 1963); V. I. Masal'skii, Ovragi chernozemnoi polosy Rossii, ikh rasprostranenie, razvitie i deiatel'nost' (St. Petersburg, 1897).

5. A. F. Guzhevaia, Ovragi Sredne-Russkoi vozvyshennosti, Trudy Istituta Geografii 42 (1948): 71.

6. K. V. Dolgopolov, Tsentral'no-chernozemnyi raion (Moscow, 1961), p. 19; G. M. Lappo, ed., Tsentral'nyi raion (Moscow, 1970), pp. 644-45.

7. G. P. Surmach, Pochvenno-erozionnye issledovaniia na srednerusskoi vozvyshennosti, Sel'skokhoziaistvennaia eroziia i bor'ba s nei (Moscow, 1956), pp. 93 and 101.

8. P. S. Zakharov, Eroziia pochv i mery bor'by s nei (Moscow, 1971), pp. 50-51.

9. S. I. Sil'vestrov, Eroziia i sevooboroty (Moscow, 1949), pp. 33-55.

10. D. L. Armand, Antropogennye erozionnye protsessy, Sel' skokhoziaistvennaia eroziia i bor'ba s nei (Moscow, 1956), pp. 8-9.

11. L. S. Berg, Natural Regions of the U.S.S.R., (New York: Macmillan, 1950), p. 84; Grin, Dinamica vodnogo . . . , p. 137; G. D. Rikhter and F. N. Mil'kov, eds., Lesostep' i step' Russkoi ravniny (Moscow, 1956) p. 131.

12. Semenov, Rossiia, Vol. 2, pp. 542-43; E. G. Shuliakovskii, ed., Ocherki istorii Voronezhskogo kraia s drevneishikh vremen do velikoi oktiabr'skoi sotsialisticheskoi revoliutsii, 2 vols (Voronezh, 1961), Vol. 1, p. 86.

13. V. K. Iatsunskii, Izmeneniia v razmeshchenii zemledeliia v Evropeiskoi Rossii s kontsa XVIII v. do pervoi mirovoi voiny, Voprosy istorii sel'skogo khoziaistva krestianstva, i revoliutsionnogo dvizheniia v Rossii (Moscow, 1961), p. 126.

14. V. K. Iatsunskii, Izmeneniia v razmeshchenii naseleniia Evropeiskoi Rossii v 1724-1916 gg. Istoriia SSSR, no. 1 (1957), pp. 203 and 210.

15. Semenov, Rossiia, Vol. 2, p. 203.

16. V. F. Nagorskii, Oskudenie skotom tsentral'nykh gubernii Evropeiskoi Rossii s 1864 po 1892 god, Trudy Orlovskago oblastnogo s'ezda sel'skikh khoziaev (Orel, 1898), p. 225.

17. Materialy vysochaishe uchrezhdennoi 16 noiabria 1901 g. Komissii po issledovaniiu voprosa o dvizhenii s 1861 po 1900 gg.

blagosostoianiia sel'skogo naseleniia srednezemledel'cheskikh gubernii (St. Petersburg, 1903).

18. D. Voeikov, Ekonomicheskoe polozhenie krestian v chernozemnykh guberniiakh Izvestiia Imperatorskago Russkago Geograficheskago Obshchestva, no. 3 (1881), p. 6.

19. P. A. Zaionchkovskii, Otmena krepostnogo prava v Rossii (Moscow, 1968), p. 240.

20. Grin, Dinamika vodnogo . . . , p. 102.

21. G. D. Rikhter and A. E. D'iachenko, eds., Sel'skokhoziaistvennaia eroziia i bor'ba s nei (Moscow, 1956), p. 3.

22. N. P. Aleksandrov, ed., Razmeshchenie i spetsializatsiia zemledeliia i zhivotnovodstva v tsentral'no-chernozemnoi zone (Moscow, 1968), p. 20.

23. Postanovlenie Soveta Ministrov Soiuza SSR i Tsentral'nogo Komiteta VKP (b) ot 20 oktiabria 1948 goda "o plane polezaschitnykh lesonasazhdenii, vnedreniia travopol'nykh sevooborotov, stroitel'stva prudov vodoemov dlia obespecheniia vysokikh i ustoichivykh urozhaev v stepnykh i lesostepnykh raionakh evropeiskoi chasti SSSR," pp. 343-44, 351-54, 359, and 372, appended in Dmitriev, O travopol'noi sisteme zemledeliia, pp. 335-72, map.

24. Dmitriev, O travopol'noi sisteme zemledeliia, pp. 5 and 10.

25. Sil'vestrov, Eroziia i sevooboroty, p. 16.

26. W. A. D. Jackson, ed., Natural Resources of the Soviet Union: Their Use and Renewal (San Francisco: Freeman, 1971), p. 161.

27. Rikhter and D'iachenko, Sel'skokhoziaistvennaiia . . . , pp. 3-4.

28. S. Ploss, Conflict and Decision-Making in Soviet Russia: A Case Study of Agricultural Policy, 1953-1963 (Princeton: Princeton University Press, 1965), pp. 67-97.

29. D. L. Armand, Izuchenie erozii v lesostepnykh i stepnykh raionakh SSSR i sostoianie protivoerozionnykh meropriiatii, Izvestiia Akademii Nauk SSSR, Seriia geograficheskaia, no. 2 (1954), pp. 11-13.

30. Ibid., p. 9.

31. Ibid., pp. 9-10.

32. Dolgopolov, Tsentral'no-chernozemnyi raion, p. 20.

33. A. I. Voeikov, Zemel'nye uluchsheniia i ikh sootnoshenie s klimatom i drugimi estestvennymi usloviiami. Ezhegodnik otdela zemel'nykh uluchshenii, Vol. 1, (1910), p. 98, reprinted in Voeikov, Vozdeistvie cheloveka na prirodu.

34. T. F. Antropov and D. L. Armand, Organizatsiia territorii i sevooboroty v kolkhozakh erodirovannykh raionov Sredne-russkoi vozvyshennosti, Sel'skokhoziaistvennaia eroziia i bor'ba s nei (Moscow, 1956), p. 247.

35. E. A. Mironova, Opyt morfometricheskoi kharakteristiki erozionnogo rel'efa, Sel'skokhoziaistvennaia eroziia i novye metody ee izucheniia (Moscow, 1958), p. 215.

36. Zakharov, Eroziia pochv i mery bor'by s nei, pp. 3-4.

5

**THE FALLING LEVEL
OF THE CASPIAN SEA**
Philip P. Micklin

The level of the Caspian Sea in the USSR, the world's largest lake, has been steadily declining for four decades (see Map 5.1). By the late 1960s, average sea level was three meters lower than it was in 1929. Pronounced, long-term level changes are not unusual for the Caspian. Historical and archeological evidence augurs that the sea's level has fluctuated at least eight meters over the last two millennia.[1] However, the recent drop is the most severe in five centuries and has occurred against the background of an industrializing society. Consequently, it has caused considerable economic dislocation and raised the utmost concern within the Soviet government and scientific community.

CLIMATIC AND HUMAN CAUSES

The major cause of the level fall is a marked reduction in river inflow, particularly from the Volga, which contributes, on the average, 78 percent of overall surface discharge to the Caspian. Changes in other elements of the water balance—ground water influx, precipitation on and evaporation from the sea's surface, and outflow to the lower lying Gulf of Kara-Bogaz-Gol—have not contributed appreciably to the decline. Explaining the reduction of river discharge is more complex. Climatic factors have obviously played a prime role. Winter precipitation over the northern Volga Basin, the chief flow-generating area for the Caspian, has been generally lower since 1930 than in previous years. This departure was most pronounced from 1930 to 1940, the period of the most rapid level drop. Altered atmospheric circulation over the European USSR has been connected with the precipitation diminution.[2]

MAP 5.1

The Caspian Sea

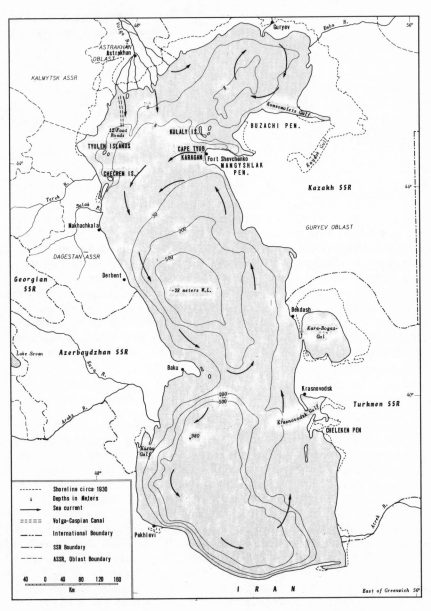

Source: Composed by the author.

But human activities have also reduced inflow to the sea, particularly over the past 25 years. Reservoir formation, irrigation, municipal and industrial withdrawals, and perhaps agricultural practices have been of major importance. Most of the water has been eliminated from the Volga where a chain of hydroelectric stations and their associated reservoirs have been erected since the late 1930s. The human-caused deficit for 1945-65 has been estimated at 451 km^3.[3] Average annual eliminations could reach 62 km^3 in 1980 and 82 km^3 by the year 2000.[4] The weight of evidence suggests that low river flow and the associated drop of sea level have been more the result of human interference than climatic factors since the mid-1950s. Sea level in the future undoubtedly will be more dependent on human actions within the Caspian Basin than on climatic alterations over it and, unless corrective measures are undertaken, it should continue to decline for at least several more decades.[5]

ECONOMIC AND ENVIRONMENTAL CONSEQUENCES

The nationally important Caspian fishery has suffered most severely from the level decline. In the early 1930s annual catches were around 600,000 metric tons, but by the late 1960s they had decreased to less than 100,000 metric tons.[6] Equally important, the mix of fish has altered radically with low-value species, chiefly sprats, composing a much greater share of the catch in recent years, whereas the take of prized types such as sturgeon, whitefish, salmon, and herring has dwindled considerably. Losses during the late 1950s were estimated at 100 million rubles annually.[7] Harm to the fishery owing to the level decline has come primarily from the disappearance of shallows and secondarily from the increased salinity of the northern Caspian. Shallows are the biologically most productive areas of the sea, providing a food base for the more valuable types of fish and also serving as a spawning ground for some species. Most of the shallows loss has occurred in the northern part of the Caspian, the area of which has shrunk almost 25 percent over the past four decades. The average salinity of the northern Caspian, owing to current changes and the decreased flow of the Volga, has increased approximately 30 percent since the 1930s.[8] This has attenuated the area environmentally favorable for some commercially valuable species.

The sea's recession has not been the sole cause of the fishery deterioration. Damming of inflowing rivers, irrational fishing practices, pollution, and irrigation have also played a role. Accurate assessment of the significance of individual factors in the overall problem is difficult, but the falling level perhaps has been the most important contributor.

Shipping has also suffered markedly from the Caspian's drop. Channels have shallowed, many wharves have become unusable, and new shoals and reefs have complicated navigation. Costs of extensive and repeated dredging, rebuilding, and relocation of port facilities, addition of shallow-draft vessels to the fleet, and the partial loading of older, larger ships have been considerable.

The level of decline has appreciably affected other economic sectors. Agriculture in areas adjacent to the sea and the Caspian oil industry have both benefited and suffered. Thus the sea's recession has complicated irrigated agriculture in the Volga Delta but has led to the uncovering of a sizable strip suitable for agriculture along the coast of Gurev oblast'. By the same measure, shallowing has complicated the operation of offshore oil wells but at the same time has exposed oil-bearing shore areas and brought formerly inaccessible sea bottom within drilling depth. Sulfate extraction based on the brines of Kara-Bogaz-Gol has definitely suffered. Diminished flow into the gulf because of the Caspian's drop led to increased salinity and the consequent precipitation of unwanted salts along with the desired sodium sulfate. Costly measures have had to be instituted to maintain production.

The fall in sea level has had perceptible environmental consequences. The most serious of these have resulted from the approximately 25 percent reduction in the area of the northern Caspian. The climate along the northern and northwestern coast of the sea has become more continental and desert conditions have intruded into the eastern portion of the Volga Delta.[9] Of more consequence, the reduced area of the northern Caspian has been related to an increased frequenc of damaging Sukhovei (hot, dry winds blowing out of the Transcaspian deserts) in extreme southeastern European Russia.[10]

The drop in sea level has unquestionably had an overall negative impact on Caspian-related economic activities. K. K. Giul of the Institute of Geography of the Azerbaidzhanian Academy of Sciences estimates aggregate losses reached 10 billion rubles by the late 1960s.[11] However, the procedure used to derive this figure is unknown Thus, sufficient allowance may not have been made for the benefits resulting from a lower sea level, particularly those that are indirect, such as the value accruing to economic activities that reduced the inflow of water to the Caspian.

CORRECTIVE MEASURES

The prolonged drop of the Caspian, the extensive economic harm it has wrought, and predictions of even lower levels in the future, have generated considerable enthusiasm for regulatory measures.

The goal is sea level stabilization to truncate or eliminate long-term cyclic fluctuations.[12] The more grandiose alleviatory schemes would entail considerable environmental alteration and could have widespread undesirable ecologic and environmental side effects that are difficult to foresee.

Complicating the institution of corrective action is sharp controversy over the best level. Some experts have believed a raising of sea level is warranted.[13] Others have insisted the modern standing or an even lower level to be most desirable.[14] A significantly higher level may appear a logical choice since the sea's recession has caused considerable economic harm. But this assumption is open to question. In the first place, new coastal facilities have been built on the premise of a continued low sea level. Raising it substantially would necessitate expensive protective measures or relocation of these installations. Second, raising the level would not necessarily restore former fishery productivity, a primary argument of advocates of a higher standing. Other factors besides the sea's decline have adversely influenced the catch and would not be mitigated by a higher level. Third, engineering schemes to raise the Caspian would be much more expensive than those aimed at maintaining a low level; the extra costs must be balanced against the expected benefits of a higher level. Lack of a definitive answer to the optimal-level question is a major roadblock to selection of rational regulation measures.

The simplest schemes to regulate the Caspian's level would involve reducing evaporation or lessening the flow into Kara-Bogaz-Gol. The former could be accomplished via separation of the shallow northeastern part of the sea by a low earthen dike and the latter by a concrete dam and sluice gate across the strait connecting the Caspian to the gulf.[15] The water savings from such measures could range from as little as 4 km^3 by reducing the flow into the gulf to as much as 37 km^3 if the northeastern Caspian were also isolated. The Caspian fishery could suffer gravely from a further marked reduction of shallows accompanying the diking project.

More complicated measures would involve separating and raising the level of only the northern Caspian without increasing inflow to the sea or massive supplementation of the sea's water balance through diversion from extrabasin sources. The rationale for separation and raising of the northern Caspian is that the majority of economic damage has resulted from the shrinkage of this portion of the sea. Assertedly, restoration of sea level to near the 1930 mark would be highly beneficial and could be realized with an annual Volga flow averaging no more than 220 km^3, which is less than the river's average flow for 1961-65 of 230 km^3. The heart of the project would be a 375 km earth and rock dike with two canals through it for the passage of fish and ships (Map 5.2). Cost of the undertaking would be at least 160 million rubles.[16]

MAP 5.2

The Apollov Plan

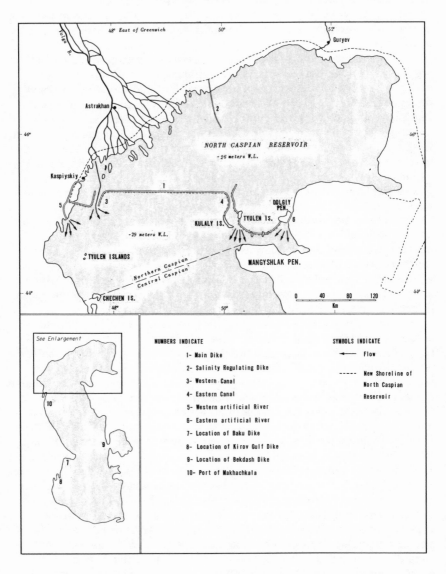

Source: Adapted from B. A. Apollov and S. N. Bobrov, "Kas-puiskoe more budet zhit," Priroda, no. 6 (1963), p. 74.

The dike could pose several problems. The level south of it would fall anywhere from 1 to 2.5 meters, requiring institution of elaborate, costly engineering measures to maintain adequate depths in the major ports of the central and southern Caspian. Habitat conditions for fish could also be worsened through radical alteration of the sea's circulation pattern, hindrance of the seasonal migration of some species, and excessive lowering of the salinity and accumulation of pollutants in the northern Caspian. Finally, the proposed dike would be fragile and could be breached during a major storm with calamitous consequences.

Three diversion proposals have been seriously considered: from the Azov Sea, from the Siberian rivers Ob and Enisei, and from several Arctic-flowing rivers of the northern European USSR. The Azov lies only 600 km from the Caspian across a low divide. Diversion could be accomplished by a gravitational canal or by pumping (see Map 5.3).[17] Cost estimates for different variants range from 600 million to over 2.5 billion rubles.[18] The Ob and Enisei lie some distance from the Caspian, but the diversion route would be across lowlands with only one significant intervening relief feature, the Turgay Gate with an elevation of 124 meters. A series of huge dams, reservoirs, and canals would be required for the project (see Map 5.4).[19] The undertaking could cost 100 billion rubles and take a half-century to complete. The most promising rivers for diversion in northern European Russia are the Pechora and Vychegda. They could be connected to the Kama, the Volga's major tributary, by a system of dams, reservoirs, and canals (see Map 5.5).[20] Cost of the project could run over a billion rubles.[21]

The quantity of water diverted by these schemes would be immense: up to 315, 250, and 42 km^3 annually from the Siberian rivers, the Azov Sea, and the Pechora and Vychegda, respectively. Each of these projects would supply more than enough water to significantly raise the Caspian's level given present water balance conditions. Besides raising the sea level, the schemes could provide other major benefits. For example, the most grandiose variant of Siberian diversion, the Davydov Plan, would in addition provide 13 million kilowatts of hydroelectric generating capacity, irrigate 20 million hectares of land in the Aral-Caspian Basin, raise and stabilize the Aral Sea, ameliorate the arid climate of Central Asia, and provide a deep-water transportation route from the Caspian to the Arctic.

Although bold in conception and impressive in scope, these projects could have serious environmental and ecological consequences. The Azov diversion could markedly increase the Caspian's salinity, adversely affecting the sea's fishery. The other two schemes would flood huge areas (up to 250,000 km^2 in the case of the Siberian diversion) and could alter the climate over extensive adjacent territories.

73

MAP 5.3

The Azov Diversion
(according to Stas)

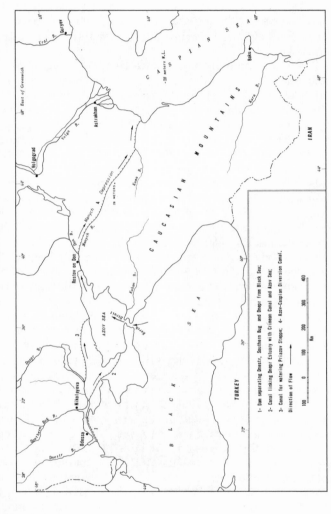

1- Dam separating Dnestr, Southern Bug and Dnepr from Black Sea;
2- Canal linking Dnepr Estuary with Crimean Canal and Azov Sea;
3- Canal for watering Priazov Steppe; 4- Azov-Caspian Diversion Canal.

Direction of Flow ➜

Source: Adapted from I. I. Stas, "Kak spasat Kaspii," Priroda, no. 12 (1968), p. 77.

74

MAP 5.4

The Davydov Plan

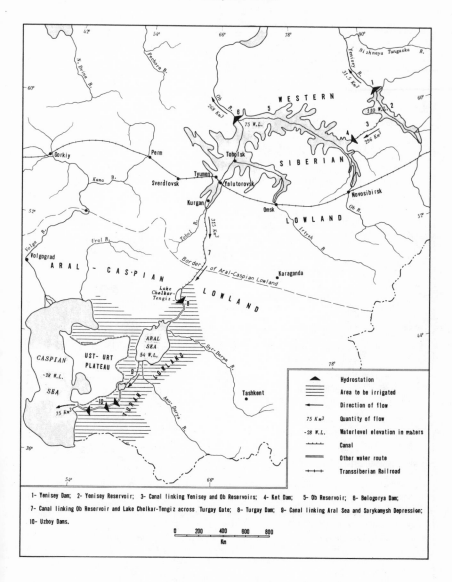

1- Yenisey Dam; 2- Yenisey Reservoir; 3- Canal linking Yenisey and Ob Reservoirs; 4- Ket Dam; 5- Ob Reservoir; 6- Belogorya Dam;
7- Canal linking Ob Reservoir and Lake Chelkar-Tengiz across Turgay Gate; 8- Turgay Dam; 9- Canal linking Aral Sea and Sarykamysh Depression;
10- Uzboy Dams.

Source: Adapted from M. M. Davydov, "Ob-Aralo-Kaspiiskoe vodnoe soedinenie," Gidrotekhnicheskoe stroitelstvo, no. 3 (1949), p. 10, and "Preobrazovanie rechnoi seti sovetskoi strany," Geografiia v shkole, no. 3 (1949), p. 15.

MAP 5.5

The Kama-Vychegda-Pechora Project
and Its Area of Influence

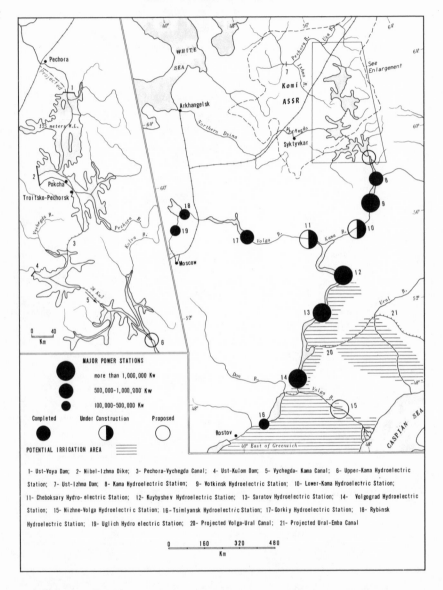

1- Ust-Voya Dam; 2- Nibel-Izhma Dike; 3- Pechora-Vychegda Canal; 4- Ust-Kulom Dam; 5- Vychegda- Kama Canal; 6- Upper-Kama Hydroelectric Station; 7- Ust-Izhma Dam; 8- Kama Hydroelectric Station; 9- Votkinsk Hydroelectric Station; 10- Lower-Kama Hydroelectric Station; 11- Cheboksary Hydro- electric Station; 12- Kuybyshev Hydroelectric Station; 13- Saratov Hydroelectric Station; 14- Volgograd Hydroelectric Station; 15- Nizhne-Volga Hydroelectric Station; 16-Tsimlyansk Hydroelectric Station; 17-Gorkiy Hydroelectric Station; 18- Rybinsk Hydroelectric Station; 19- Uglich Hydro electric Station; 20- Projected Volga-Ural Canal; 21- Projected Ural-Emba Canal

Source: G. L. Russo, "Problema ratsionalnogo ispolzovaniia stoka severnykh rek," Gidrotekhnicheskoe stroitelstvo, no. 7 (1961), p. 13.

Damming the various rivers would block the spawning run of valuable anadromous fish whereas diverting a substantial part of these rivers' flow would probably undesirably alter ecological conditions for marine biota in their estuaries. Indeed, some research has indicated the Siberian diversion would so reduce fresh water discharge to the Kara Sea that the exchange of water between the Atlantic and Arctic oceans would be drastically reduced.[22] This could lead to marked cooling of the Arctic, owing to the lessened influx of warm Atlantic water, and perhaps to climatic changes affecting the entire Northern Hemisphere.[23]

CONCLUSION

The modern recession of the Caspian has caused considerable economic harm. The drop initially owed to climatic influences, but human flow-reducing actions have assumed principal importance in recent decades. A variety of engineering schemes have been proposed to lessen future economic damage through regulation of the Caspian's level. Institution of corrective measures, however, is complicated by lack of agreement on the optimal level for the sea. The grander projects not only would be enormously expensive but long in construction. That the predicted, but not universally accepted, benefits justify such expenditures and that social and economic needs can be forecast decades into the future is debatable. More importantly, implementation of these measures could result in widespread, and perhaps irreversible, environmental and ecological damage.

Unfortunately, scant attention has been given to less environmentally disruptive methods of coping with the situation such as moderation of flow-diminishing pursuits in the Caspian Basin or adjustment of sea-related economic activities to fluctuations of sea level. On the contrary, a basic premise has been and remains that the Caspian's level must be regulated, via massive construction works, to facilitate economic development. This reflects an optimism prevalent in all industrial-technological societies that man not only possess the technical capability but the requisite wisdom to turn the environment to his advantage.

NOTES

1. B. A. Apollov, Kaspiiskoe more i ego bassein (Moscow: Akademiia Nauk, 1956), pp. 5-16, 67-88; Institut okeanologii, Trudy, Vol. XV: Kolebaniia urovnia Kaspiiskogo moria (Moscow: Akademiia Nauk, 1956), pp. 72-229.

2. A. N. Afanasev, Kolebaniia gidrometeorologicheskogo rezhima na territorii SSSR (Moscow: "Nauka," 1967), pp. 188-98.

3. G. P. Kalinin, K. I. Smirnova, and O. I. Sheremetevskaia, "Vodnobalansovye raschety budushchikh urovnei Kaspiiskogo moria," Meteorologiia i gidrologiia, no. 9 (1968), p. 47.

4. Ibid., p. 49.

5. B. A. Apollov and S. N. Bobrov, "Kaspiiskoe more budet zhit," Priroda, no. 6 (1963), p. 72; Kalinin et al., "Vodnobalansovye . . . ," p. 51.

6. I. I. Stas, "Kaspiiskuiu problemi neobkhodimo reshat kompleksno," Okeanologiia, no. 2 (1964), p. 355; A. F. Antonnikov, "Gidrostroitelstvo i rybnye khoziaistvo," Priroda, no. 9 (1970), pp. 4-5.

7. V. I. Popov, "Problema regulirovaniia urovnia Kaspiiskogo moria," in Vsesoiuznoe soveshchanie po probleme Kaspiiskogo moria (Moskva, 11-16 Aprelia, 1960), materialy (Baku: Akademiia Nauk Azerbaidzhanskoi SSR, 1963), p. 336.

8. L. G. Vinogradov and E. A. Iablonskaia, "Problemy rybokhoziaistvennoi melioratsii Kaspiiskogo moria," in Vsesoiuznoe gidrobiologicheskoe obshchestvo, Akademiia Nauk SSSR, Izmenenie biologicheskikh kompleksov Kaspiiskogo moria poslednie desiatiletiia (Moscow: "Nauka," 1965), p. 9.

9. Apollov and Bobrov, "Kaspiiskoe more budet zhit," p. 70.

10. Apollov, Kaspiiskoe more i ego bassein, pp. 40-41.

11. M. K. Grave, "Kakim dolzhen byt uroven Kaspiia?" Izvestiia Akademii Nauk SSSR, seriia geograficheskaia, no. 4 (1970), p. 159.

12. Institut okeanologii, Trudy, Vol. XV, p. 28.

13. B. A. Apollov, "Problema Kaspiiskogo moria i eo osnovnye zadachi," in Vsesoiuznoe soveshchanie po probleme Kaspiiskogo moria (Moskva, 11-16 Aprelia, 1960), materialy, p. 10; Grave, "Kakim dolzhen byt uroven Kaspiia?," p. 159.

14. S. L. Vendrov and S. Iu. Geller, "Geograficheskie aspekty Volvo-Kaspiiskoi problemy," in Sovremennye problemy geografii (Moscow: "Nauka," 1964), pp. 338-39.

15. E. M. Kopaigorodskii, "Izmenenie vodnogo balansa Kaspiiskogo moria putem sokrashcheniia ploshchadi ispareniia," Okeanologiia 8, no. 1 (1967): pp. 116-119; P. V. Nikolaeva, "Kratkii obzor skhem predlozhenii po stabilizatsii i regulironaviiu urovnia Kaspiiskogo moria," in Okeanograficheskaia komissiia, Trudy, Vol. V: Problemy Kaspiiskogo moria (Moscow: Akademiia Nauk, 1959), pp. 58-59.

16. B. A. Apollov, "Kaspiiskaia problema i puti eo razresheniia," in Okeanograficheskaia komissiia, Trudy, Vol. V: Problemy Kaspiiskogo moria.

17. I. I. Stas, "Kak spasat Kaspii," Priroda, no. 12 (1968), pp. 76-79.

18. Nikolaeva, "Kratkii obzor . . . ," pp. 54-55.

19. M. M. Davydov, "Ob-Aralo-Kaspiiskoe vodnoe soedinenie," Gidrotekhnicheskoe stroitelstvo, no. 3 (1949), pp. 6-11, and "Preobrazovanie rechnoi seti sovetskoi strany," Geografiia v shkole, no. 3 (1949), pp. 12-18.

20. G. L. Sarukhanov, "Pechora-Kaspii, reki severa potekut na iug," Priroda, no. 7 (1961), pp. 53-57.

21. Komi filial, Akademiia Nauk SSSR, O vliianii perebroski stoka severnykh rek v bassein Kaspiia na narodnoe khoziaistvo Komi ASSR (Leningrad: Akademiia Nauk, 1967), pp. 191-92.

22. V. S. Antonov, "Problema urovnia Kaspiiskogo moria i stok severnykh rek," in Trudy Articheskogo i Antarkticheskogo nauchno-issledovatelskogo instituta, no. 253 (Moscow: Akademiia Nauk, 1963), pp. 233-40.

23. Hubert Lamb, "USSR Plan for Rivers Called Risk," Detroit Free Press, February 24, 1970, p. 5C.

THE LAKE BAIKAL CONTROVERSY:
A SERIOUS WATER POLLUTION THREAT
OR A TURNING POINT IN
SOVIET ENVIRONMENTAL CONSCIOUSNESS
Craig ZumBrunnen

Until recently, Soviet and other socialist economists and theo-
rists have argued that a socialist organization of society precludes
environmental problems such as air and water pollution. However,
as with most social and economic theories, a substantial disparity
exists between theory and reality. The resolution of the great economic
debates concerning the tempo and strategy of economic growth and
development that took place in the Soviet Union during the latter 1920s
culminated in the birth of an ambitious, determined, austere development policy. Since then, the Soviet leadership has pursued aggres-
sively and quite successfully a strategy of accelerated economic growth
In the process an elaborate system of success indicators has evolved
that has placed highest priority upon production output. Hence, at
least until recently, questions regarding environmental quality have
been relegated to a quite subservient position. As a consequence,
many of the Soviet Union's waterways are presently being polluted
to varying degrees by industrial and domestic sewage. One of these
waterways is Lake Baikal.
    With respect to environmental pollution problems, the Lake
Baikal controversy is significant for several reasons. First, this
controversy is important because it has received perhaps far more
publicity within the Soviet Union than any other water pollution problem area, including the Volga River Basin. The obvious reason for
this amount of publicity is simply the sheer scale of the controversy,
which has involved thousands of Soviet officials, scientists, writers,
poets, planners, and ordinary citizens. This is surprising to West-
erners who tend to regard the Soviet Union as a closed society where
publicized conflicts such as the Lake Baikal controversy usually represent interagency rivalries. The Baikal conflict, in fact, has the
ingredients of a genuine public protest or outcry similar to such

conservation struggles in the United States as the fight to prevent construction of the Alaskan oil pipeline or forestall additional hydroelectric developments in the Grand Canyon. Accordingly, this controversy has played an important role in the evolution of official Soviet attitudes and policies toward the physical environment.

At the same time the Baikal Basin has clearly received more attention by American scholars and journalists than any other Soviet environmental problem area. Philip Micklin, Theodore Shabad, Philip Pryde, Marshall Goldman, and Victor L. Mote have discussed in various contexts the water quality problems of Lake Baikal.[1]

Both the scale of the conflict and the worldwide uniqueness of the Baikal Basin have generated international interest. The purpose of this chapter, in addition to reviewing and updating the conflict, is to depict the absolute and relative magnitude of present and probable future water pollution problems of the Baikal Basin. In this respect, the chapter takes partial exception to the alarmist tone of Goldman's three publications. While the author is in substantive agreement with many of Goldman's arguments that the Soviet command economy has built-in incentives to pollute, just as the capitalist system has, the author considers Goldman to have overstated the current pollution threat to Lake Baikal. Compared with many of the rivers of European Russia, particularly in the Volga Basin and the southeastern Ukraine, both Baikal's absolute and relative pollution levels are minor. Hence, the subtitle of Goldman's article in Science ("From Lake Erie to Lake Baikal, Los Angeles to Tbilisi the debates and dilemmas are the same") may be misleading. While the debates and dilemmas may well be the same, certainly the levels of water pollution in Lake Baikal and air pollution at Tbilisi are far less than those in Lake Erie and at Los Angeles, respectively. In fairness, the subtitle may be the words of Science's editors rather than Goldman. Regardless, Lake Superior or California's Lake Tahoe constitute a much more meaningful analogy.

In fact, this analogy has recently received official recognition by the September 21, 1972, memorandum implementing the previously mentioned May 23, 1972, Soviet-American accord signed in Moscow by Presidents Podgornyi and Nixon.[2] Among the initial projects of the cooperative agreement is a joint project on water pollution with Lake Baikal, Lake Tahoe, and one of the Great Lakes designated as the study areas. This development indicates that the Lake Baikal controversy will remain a focus of worldwide attention for some time to come.

THE GEOGRAPHICAL SETTING

As shown in Map 6.1, Lake Baikal is located in southeastern Siberia. This sickle-shaped waterbody stretches for 636 km from

81

MAP 6.1

Lake Baikal Region

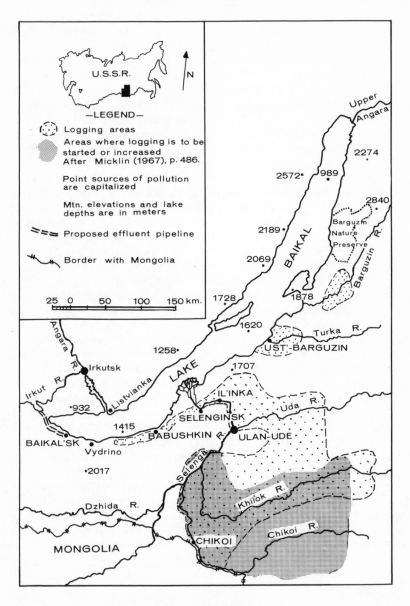

Source: Composed by the author.

its southeastern end, which is only slightly more than 100 km from the Mongolian People's Republic, to its northeastern extremity. Its width averages 47.8 km and reaches a maximum of 79.4 km. The average elevation of its surface is 455 meters above sea level. Having a surface area equal to 31,500 km^2, Lake Baikal represents the third largest "lake" by area in the Soviet Union, being exceeded only by the Caspian and Aral seas.[3] Formed by ancient and on-going tectonic displacements, the deep lake depression is bound on the northwest and on the southeast by steep mountain chains that parallel its shore-line. The peaks of these mountains tower anywhere from 500 to 2,400 meters above the water level of the lake.[4] Although data on the maximum depth of the lake vary in different references, the maximum is at least 1,620 meters, making Baikal the deepest lake in the world.[5]

Even more significant than its depth is its volume. With a volume of 23,600 km^3, Lake Baikal is the largest fresh-water lake in the world. This is approximately equal to two-thirds of the annual flow of all the rivers of the world.[6] Various individuals claim that Baikal accounts for anywhere from 2.5 to 20 percent of the world's total supply of fresh water.[7] This wide discrepancy probably is attributable to different definitions of "total supply of fresh water." In any case Lake Baikal contains approximately 18.2 percent of the world's total fresh water contained in fresh-water lakes and rivers (at any one time.)[8]

A total of 336 rivers draining an area of 557,000 km^2 flow into the lake along its 2,100 km shoreline (see Map 6.2). The watershed of the largest river, the Selenga, constitutes 447,060 km^2 or 82.8 percent of the entire Baikal watershed, and supplies about 51 percent of the lake's water. Other smaller, yet major tributaries include the Barguzin, Turka, Goloustnaia, and the Upper Angara.[9] By contrast, the lake is drained by a single river, the Angara. Flowing out of the southwest end of the lake, the Angara joins the Enisei 1,779 km to the northwest. The Enisei, in turn, flows 2,000 km north before discharging into Enisei Bay in the Arctic Ocean Basin.[10]

The most significant and relevant characteristic for the stated purpose of this chapter is the world-famous purity of Baikal's water. A large proportion of the lake's watershed consists of Pre-Cambrian and Paleozoic granitic intrusives overlain by thin forest soils.[11] Because granitic rocks tend to be relatively resistant to mechanical and chemical weathering and because the severe climate of the Baikal area retards chemical weathering processes, the rivers feeding Baikal for millennia have been discharging water containing only minor quantities of both dissolved and suspended substances. Hence, the lake water is transparent to an average depth of 26 meters with a maximum depth of about 40 meters.[12] The low silica content (SiO_2) ranges from 1.6 to 5.5 mg/1 while the average total mineralization

MAP 6.2

The Lake Baikal Watershed

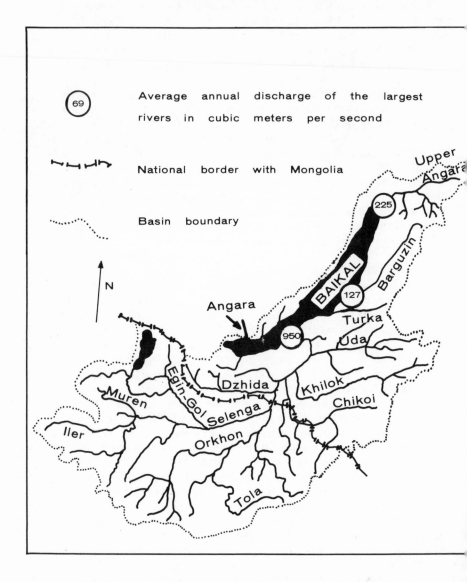

Average annual discharge of the largest rivers in cubic meters per second

National border with Mongolia

Basin boundary

Source: L. Rossolimo, Baikal (Moscow: Izdatel'stvo "Nauka 1966), p. 62.

is about 100 mg/1.[13] For comparison, Lake Superior, which has a drainage area mainly underlain by Pre-Cambrian shield, averages 81.7 mg/1 whereas Lake Michigan averages 167.4 mg/1 in total mineralization.[14] Furthermore, the lake water has a very low content of organic substances. Finally, the lake is richly endowed with dissolved oxygen averaging 12-14 mg O_2/1.[15]

One final fact should be mentioned before discussing the water quality problems of Lake Baikal. Partly as a result of the purity and great antiquity of the lake (not less than 26-30 million years old), Baikal is the habitat for over 1,800 species of flora and fauna, three-fourths of which are endemic to Baikal. Included among the fauna are several varieties of commercially valuable fish, the most important of which is the omul.[16] Thus, in addition to being the largest natural factory in the world for the production and storage of fresh water, the lake is a living museum of the past.

However, the pristine quality of Baikal's water and the unique assemblage of organic life dwelling in it are being threatened directly and indirectly by the actions of industrial man.

THE THREAT FROM INDUSTRIAL ENCROACHMENT

Before the 1960s there was little significant industrial development in the Baikal Basin in general and on the lake in particular. However, with the Soviet Union's determined drive to develop the natural resource wealth of Siberia, it is not surprising that planners and industrial managers began to view the water and timber of the Lake Baikal Basin as greatly underutilized resources. For a long time the forestry industry, the main industry in the area, had been developed in a lopsided fashion; that is, no wood-processing enterprises existed to complement the intensive lumber procurement activities. Thus, in an attempt to integrate more fully the forest resources of the area, the State Institute for the Design of Cellulose and Paper Plants, with the encouragement of the Russian Republic State Economic Planning Commission (Gosplan), suggested in 1957 that the vast resources of the Baikal Basin be utilized. The plans announced in 1958 called for the construction of a large cellulose plant on the lake's major tributary river, the Selenga.[17]

PROTAGONISTS INVOLVED IN THE CONTROVERSY

The building of these two plants and their possible effects upon the water quality of the lake and its biota became the major focus of contention in the environmental conflict that has evolved (see Map 6.1).

As in conservation battles in the United States, the lines have been drawn between those groups interested in economic development and those concerned more with the protection and preservation of the more rare and beautiful aspects of nature. In this debate the economic interests have been most strongly represented by the East Siberian Economic Council,[18] the Irkutsk Economic Council,[19] the State Committee for the Lumber, Pulp and Paper and Wood-Processing Industry and Forestry, and the State Institute for the Design of Pulp and Paper Industry Enterprises in Siberia and the Far East (Sibgiprobum).[20] Their conservationist opposition has been comprised of a somewhat diverse assemblage of leading Soviet scientists, academicians, public health officials, representatives of the fishing industry, Soviet writers and journalists, plus several other organizations and private individuals. Counted among this group have been Petr Kapitsa, the famous physicist; Boris P. Konstantinov, vice-president of the Soviet Academy of Sciences; Innokenty Gerasimov, a prominent geographer and soil scientist;[21] Iurii Danilov, USSR Deputy Minister of Public Health and USSR Chief Sanitary Physician;[22] Mikhail Sholokhov, the famous novelist,[23] and Oleg Volkov, the Soviet writer.[24]

THE BEGINNING OF THE CONTROVERSY

In February 1960, Nikolai T. Suprunenko, the director of the future Baikal'sk Cellulose Plant, visited the site that is located about 35 km southeast of Sliudianki along the south shore where the small river, Solzan, empties into Lake Baikal.[25] By June 1960, construction of the plant with a planned annual capacity of over 700,000 m^3 of lumber and 100,000 tons of sulfate cellulose was under way.[26] A month later the Buriat Book Publishing Company of Ulan-Ude published a limited edition (2,500 copies) of a cautionary essay by a local writer, B. R. Buiantuev.[27]

Although it is unlikely this minor publication had much impact, an August 1960 Izvestiia construction progress report claimed that in spite of the tens of thousands of tons of high-quality cellulose, paper, fodder yeast, and millions of cubic meters of lumber to be produced annually by the Baikal'sk plant, the Baikal taiga forest would not suffer since replacement planting would be provided.[28] Another report in September 1960 appeared pleased with the construction tempo of the building trust—Irkutskpromstroi.[29]

New warnings appeared in a minor publication from Ulan-Ude in 1961 by two local writers named S. Sarkisian and O. Serova.[30] However, the potential threat to Lake Baikal seems to have been first brought to national attention by Gregorii Galazii, director of the Limnological Institute of the Academy of Sciences, which is located

at Listvianka on the north shore of Baikal near the source of the
Angara River. In December 1961, in a letter to the editor of the newspaper Komsomolskaia pravda, Galazii set forth the development of
the plan for industrialization along the lake shore, listed the chemical
and biological changes that the new industries would render upon the
lake water, and gave warning that the plants' effluents could destroy
some of the lake's unique marine life as well as possibly adversely
affect Irkutsk's water supply.31

The growing controversy between the industrial development
interests and the cautious scientists was mentioned in Izvestiia in
July and August 1962.32 Then in 1965 the Baikal conflict became a
full-blown debate with a stream of polemic articles appearing in
Literaturnaia gazeta, Pravda, Izvestiia, and Komsomolskaia pravda.33
Although recently the furor has abated, Literaturnaia gazeta in particular continues to publish articles on the fate of Lake Baikal. Philip
Micklin appears to have been the first American to analyze this debate
and Marshall Goldman has chronicled it in The New Yorker and in
his book The Spoils of Progress.34

THE CONSERVATIONISTS' ARGUMENTS

It now seems timely to present and analyze the arguments that
have been put forward by the two groups of adversaries. The conservationists' arguments will be discussed first. Basically, the conservationists' case focuses upon two points: (1) the direct water
pollution threat they believe the Baikal'sk and Selenga plants would
pose to the lake, and (2) the indirect, adverse effects upon the lake
(and the entire basin) they believe would be the inevitable consequences
of the increased logging required to satisfy the raw material requirements of the two plants.

The Pollution Threat from Industrial Effluents

Of most direct concern to the nature protection forces has been
the fear that the two plants would discharge their effluents into the
lake and the Selenga River without purification. Mikhail Mikhailovich
Odintsov, a doctor of geology-mineralogy associated with the East
Siberian office of the Siberian branch of the Academy of Sciences of
the USSR, claimed that "the designers did not take care of the purifying
structures." According to Odintsov, "This is the destruction of the
unique lake!"35

Second, the conservationists have questioned the efficacy of the
proposed purification processes. The design for the purification

facilities at Baikal'sk was claimed to be new and hence never to have been tested under industrial operating conditions. The advanced tests were limited to a semilaboratory type of device used to purify 200 liters of artificially polluted water.[36] Then too, the development antagonists have feared that the plants would begin operation without the purification facilities completed.[37]

The Baikal'sk plant was expected to demand approximately 250,000 m^3 per day of lake water and, at the same time, to discharge directly into the lake about 220,000 m^3 per day of effluent. The Selenga plant was expected to discharge between 18,000-45,000 m^3 per day of waste water into the Selenga River. Both enterprises were to employ the sulfate process to produce cellulose.[38] Hence, the main pollutant in the effluent would be biodegradable lignin or organic wood fibers. Other organic contaminants would include wood sugars, resins and mercaptans (R-SH).

Upon discharge these organic wastes immediately begin to be broken down by aquatic biota into harmless substances. Besides the claim that the minute cellulose fibers cover the gills of fish and suffocate them,[39] the real threat of such organic pollutants is their high biochemical oxygen demand (BOD). In other words, in the "self-purification" process dissolved oxygen (DO) required for survival by most kinds of aquatic life is consumed. Thus, if the BOD is sufficiently high, fish and other organisms may suffocate.

Unfortunately, the effluent would also include inorganic substances, such as alkalies (NaOH), calcium bisulfate (Ca $[HSO_4]_2$), sodium sulfate (Na_2SO_4), and chlorine, which in sufficient concentrations are toxic to aquatic life. Since these substances are not biodegradable, there is the possibility they could build up to toxic concentrations.

Gregorii Galazii argues that it takes Baikal 400 years to renew its water and, therefore, the lake would not be able to oxidize the plants' sewage. According to his calculation the Baikal'sk plant's discharge of approximately 225,000 m^3 per day of waste water (approximately 8,000,000 ft^3 per day) would require 1 million m^3 per day (approximately 35 million ft^3 per day) of Baikal's water to oxidize it. The lake would not be capable of this level of self-purification, Galazii concludes, and the result is likely to be the gradual spread of a dead zone.[40] However, simple arithmetic reveals that even without any reoxygenation of the lake's waters, it would take approximately 63,000 years for the Baikal'sk plant to consume all the dissolved oxygen in the lake.[41] In Galazii's terms (even without any allowance for water recharge or reoxygenation), it would take 63,000 years for the "dead zone" to engulf the entire lake. On the other hand, as Figure 6.1 shows, the DO concentration decreases with depth and consequently a much higher dilution factor than 4 or 5 probably would be required

FIGURE 6.1

Vertical Distribution of Dissolved Oxygen
in Lake Baikal, Spring 1950

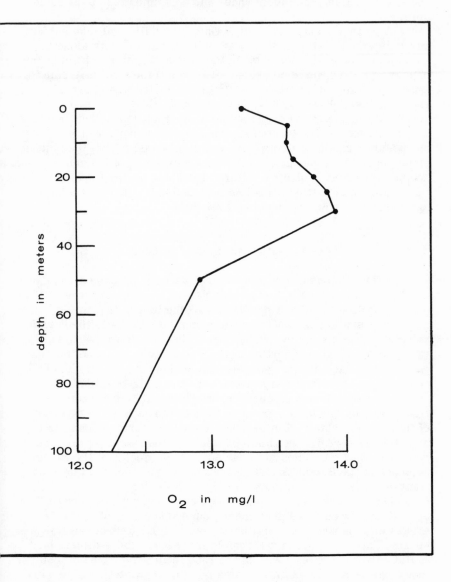

Source: L. Rossolimo, _Baikal_ (Moscow: Izdatel'stvo "Nauka,"
1966), p. 98.

if the water used were from great depths. Second, little intermixing occurs between the water above and below the 200 to 250-meter level, and it seems likely that local anaerobic contamination zones could develop.

More important, though, was Galazii's claim that lake surface currents would contaminate a zone (if one existed) north along the shallows of the eastern shore. In the process, spawning beds at the mouths of rivers would be sealed off. Even if the fish were able to spawn, the hatched small-fry would be decimated when they tried to enter the lake.[42] Figure 6.2, which schematically represents the counterclockwise gyres of the surface currents, lends credence to Galazii's hypothesis. Of course, the same hypothesis would apply to any possible contaminated waters entering the lake from the Selenga plant. In addition, because a very substantial part of the aquatic biota's habitat is located in shallow waters around the perimeter of the lake, fish spawning activities cannot be assumed to be the only biological processes that could be adversely affected.

The Threats from Increased Logging

Antidevelopment forces have also waged their campaign against the direct and indirect effects of intensified logging operations. These groups mention the direct scenic and aesthetic loss that would occur with increased logging and emphasize more specifically the harmful indirect consequences of logging that include (1) deterioration of the water conservation properties of the watershed, (2) disruption of river regimens, (3) accelerated soil erosion, and (4) changes in the chemical composition of the tributary rivers and, hence, Lake Baikal.

According to one calculation, an additional 50,000 hectares of timber would have to be harvested annually to satisfy the demands of the two pulp plants. But because of the high surface runoff of the mountainous terrain, the thin top soils, and the extremely long regeneration period, scientists conclude that even without the timber demands of the Baikal'sk and Selenga mills, the maximum advisable limit of forest utilization has already been exceeded.[43]

The conservationist Oleg Volkov claims that this amount of deforestation would result in more than 400,000 tons of soil being eroded away annually in contrast to some 60,000 tons by natural processes, thus resulting in a sevenfold increase of the sediment load in Baikal's tributaries. Already in 1966 the average annual sediment index was as high as 50 milligrams per liter or five times the permissible norm for drinking water.[44] Although this figure is relatively low compared to rivers elsewhere, the increase in suspended and dissolved solids could possibly alter the turbidity and chemical

FIGURE 6.2

Scheme of Permanent Surface Currents on Lake Baikal

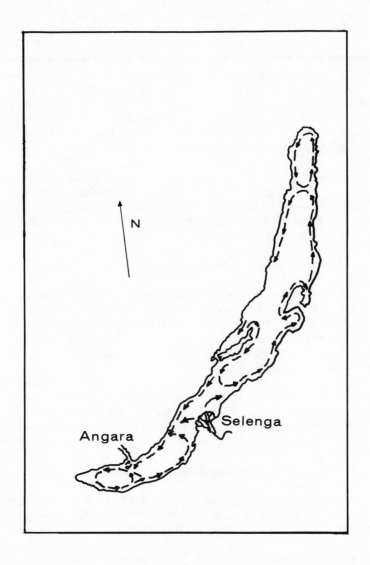

Source: L. Rossolimo, Baikal (Moscow: Izdatel'stvo "Nauka,"
1966), p. 68.

composition of the lake water enough, at least locally, to be harmful to components of Baikal's unique biota.

Volkov's assertion is supported by the Geographical Society of the USSR's 1962 statement that stresses that "cutting of forests in the Baikal area by modern methods under mountain conditions where the soil is thin will lead to destruction and erosion."[45] According to Micklin the situation is worsened further because much of the cutting is to be carried out in the Chikoi and Khilok River Basins (see Map 6.1), which receive a meager 20 to 30 centimeters of precipitation annually— quite possibly an amount insufficient to allow regrowth.[46] The geographer Shipunov strengthens this argument with logging examples from the West Siberian Economic Region. He predicts that tree-cutting would disrupt the river regimes within the affected areas by increasing the tendencies both for the flooding and for the drying up of some rivers.[47]

It is also feared that large-scale denudation resulting from increased forest harvesting could precipitate the expansion of the adjacent Gobi Desert into southern Siberia. In 1963 a scientific-production conference on soil erosion in the Buriat Autonomous Republic stated that

there are at present in the republic more than 1,500 centers, points, and areas of shifting or poorly anchored sands. . . . As a result of the movement of these sands and the development of water and wind erosion in the Buriat Autonomous Republic, 115 inhabited localities, 133 rivers, sections of 67 highways and dirt roads, and six lakes are being affected and about 20 km of irrigation canals silt up each year.[48]

Time will test the validity of the conservationists' position that current logging practices and policies will, if continued, adversely affect the natural and human environment of the Baikal Basin.

The Threat from Timber Rafting

Clearly voiced has been the complaint about particular logging techniques. The technique most severely criticized has been the transporting of timber by rafting or floating down rivers and upon the lake because about 10 percent of all the timber moved this way becomes waterlogged and sinks.[49] Consequently, the sunken logs clutter river beds, consume DO as they decay, and effectively destroy the fish spawning beds of rivers. Clearly, this method of timber transport indirectly leads to biological pollution, and if the BOD of the sunken logs becomes too high, anaerobic zones may develop.

ARGUMENTS OF THE PROPONENTS OF
INDUSTRIAL DEVELOPMENT

Proponents of industrial development have also been active in promoting their cause. Their case for pulp mills hinges mainly upon three lines of argumentation: (1) the economy's need for pulp, (2) the suitability of the two chosen sites, and (3) the claim that the water of Lake Baikal Basin and surrounding forests would not be damaged by the mills themselves.

The Need for Pulp

A first line of argument stated the Soviet Union's continued need for large quantities of pulp for the manufacture of artificial fibers and paper products. The construction promoters stressed the vital importance of artificial fiber in the manufacture of tire cord, especially for aircraft tires. In 1965 the pulpmill interests extrapolated that 347,000 tons of pulp would be required annually by 1970 to meet the demand for fiber production alone, and claimed the indispensability of the Baikal'sk mill's 200,000-ton annual capacity for high-quality viscose pulp. To further support this argument, mill construction advocates pointed out the scarcity of pulp in the USSR in 1965 and emphasized the country's need to import this strategic commodity at high prices.[50]

The Appropriateness of the Sites

Proponents of the mill construction offered two important reasons to justify their selection of the Baikal'sk and Selenga sites. First, they stressed the requirement of large volumes of ultra-pure water for the production at the Baikal'sk mill of high-quality viscose pulp. Second, they maintained that the abundant supply of pine and larch in the Baikal Basin would provide the best raw materials for the factories.[51]

Assurances of Adequate Pollution Control Measures

Early in the controversy the conservationists focused upon the potential pollution threat of the mills to the lake and upon the detrimental manifestations of increased logging, and the proponents of development have been obliged to respond by stressing that "Baikal will not be done any harm, not only in regard to preservation of its unique reservoir of fresh water with its fauna and flora, but also in regard to the cultural-aesthetic significance of the lake."[52]

To this end the State Timber Industry Committee offered assurances regarding the lake's safety, "since industrial waste waters will be purified to drinking water standards."[53] Elaborate purification installations were to be installed at each plant whereby all effluent would undergo mechanical, chemical, and biological purification followed by aeration and a final sand-filtering before being discharged. At Baikal'sk the outfall was planned for a depth of 130 meters.[54] Rem Latypov, the chief mechanic at the Baikal'sk plant, claimed that the designers had taken pollution problems into account. Waste products were to be utilized in the following manner: (1) the manufacture of fodder yeast from wood sugars, (2) the burning of bark for fuel, and (3) the conversion of resins into oils and soap.[55] Finally, a decision was made to build a 35-million-ruble sewage line from Baikal'sk to the Irkut River, a distance of approximately 70 kilometers, to divert the effluent if it were found to be harmful to the lake (see Map 6.1).[56]

Assurances of Rational Timber Procurement and Logging Techniques

To strengthen its case, the pro-pulp group was able to cite the provisions of the May 9, 1960, decree of the Council of Ministers of RSFSR designed to protect the lake. This law stipulates that (1) no industry is to begin production until its executives can guarantee the proper completion and effective functioning of the purification facilities; (2) this condition is to be certified by a state sanitary inspector and by various ministries; (3) the Barguzin Nature Preserve is to be expanded (see Map 6.1); (4) clear-cutting of timber on slopes over 15° is banned, but selective cutting on slopes over 25° is allowed; and (5) a reforestation program is to be provided.[57] The timber interests made assurances that the following additional protective measures would be taken: (1) no cutting on slopes exceeding 25°, or slopes of 15-25° if soil were thin, along small river routes, and on water divides; (2) cutting completely banned within 5 to 9 km and in some instances 30 km of the lake and in a strip paralleling both sides of the major tributary rivers; and (3) no floating of logs down small rivers, instead trucks would be utilized.[58]

THE CONSERVATIONISTS' COUNTERATTACK

Criticism of the Need for Pulp

Despite all assurances and proposed protective measures, the conservationist forces counterattacked vigorously. First, they challenged the necessity of increased pulp production. Quite rightly, they pointed out that a major demand of viscose pulp, viscose cord for tires, could be eliminated by substitution of cord synthesized from petroleum.[59] Using the United States and Japan as examples, they demonstrated that nylon was rapidly replacing cellulose cord in conventional tires and that synthetic polyesters and polyamide fibers were replacing cellulose cord in radial tires.[60]

Criticism of the Need for Baikal's Pure Water

While the conservationists conceded the Baikal'sk plant's need for pure water, if it were to make high-quality viscose pulp, they also correctly maintained that the Selenga mill did not require pure water.[61] On the contrary, the Selenga cellulose and cardboard combine scheduled to produce low-grade cellulose and cardboard products requires neither particularly pure water nor any particular species of tree.[62] Goldman cites the May 11, 1966, Komsomolskaia pravda editorial[63] which further weakens the arguments of the Ministry of Timber, Paper, and Woodworking by noting that in the mid-1960s the ministry had in fact changed the kind of cellulose to be produced at Baikal'sk and also abandoned the idea of processing pulp at the mill.[64] The plant now was also to manufacture paper not requiring exceptionally pure water. Similarly, it was only after construction had commenced on the Selenga cellulose combine that it was decided to produce cardboard.[65]

The Seismic Hazard

Many scientists have criticized the two sites, and in fact the construction of any large industrial complex in the area, on the grounds that the area is very active seismically. The Baikal'sk construction site juxtaposes an active fault. Moreover, the Selenga combine is located very near the epicenter of a relatively recent major quake.[66] Thus, it seems on both a macro- and a micro-basis that the two projects are located in active and hazardous seismic zones.[67] But to use high-seismicity as an argument to preclude industrial development in the Baikal Basin is also to make a case against continued industrial development in the Caucasus, Central Asia, and Kamchatka.[68]

Adequacy of the Raw Material Supply and Cost Override

On an issue that is still being contested, the conservationists stated that the Ministry of Timber, Paper, and Woodworking greatly overestimated the forest inventory and regeneration capacity. According to conservationists, raw material supply is only sufficient for at best 25 years against the 100 years' forecast of the ministry. At the same time critics insisted that the Ministry of Timber, Paper, and Woodworking deliberately underestimated construction costs by as much as 22 million rubles, one-third of the total investment, just to obtain initial approval for the project.[69]

Additional Fears about the Pollution of Lake Baikal

More important, however, has been the critics' continued lack of faith in the effectiveness of sewage purification facilities. Underpinning their misgivings is a 1962 report by the State Committee for the Coordination on Scientific Research. The report makes the claims that (1) the purification equipment had only been tested under laboratory, not industrial, conditions (as was mentioned previously); (2) the basic reagent for the second state of treatment was unavailable; (3) activated sludge, imperative for the biological purification process, can only be prepared at temperatures above $0°C$, a temperature encountered very infrequently in the Baikal area from October through April; and (4) the biological treatment section requires heated water and constant temperatures to function effectively, conditions that are exceedingly difficult to maintain during the winter season. As a consequence of the above considerations, the report concludes that "pollution of the coastal zone of Lake Baikal in the region of the entrance of effluent from the Baikal'sk is inevitable even with the obtainment of practical and acceptable methods of water purification."[70]

The location of the Baikal'sk plant's water intake 3.5 kilometers from the effluent outfall exemplifies the complexity of the issue.[71] Calculations made in 1965-66 by the Institute of Thermal Physics and the Computer Center of the Siberian branch of the Academy of Sciences revealed that the water intake would be in the zone of pollution from 30-35 percent of the year and in some years up to 60-66 percent of the time. Because the mill would be polluted by its own industrial wastes, additional expenditures may be required for preliminary water treatment. This issue further undermines the pure water argument used to justify the plant's initial location.

Finally, the opponents of the Baikal'sk mill doubted that purification installations would be ready by the time the plant was operative.[72] This fear was predicated upon the fact that wherever else pulp enterprises have appeared, harmful effects have resulted.

I hope the designers, builders and operators of the pulp-and-paper mills, who have not yet learned to show genuine concern for the future, will not take umbrage when I say that wherever such enterprises have appeared—on the Vychegda or in Kalinin Province, on Lake Onega, in Balakhna or Krasnoiarsk—they have caused and continue to cause a great deal of damage to living nature. There is not a single pulp-and-paper mill in our country whose waste-purification system is functioning at full capacity.[73]

Concrete Proposals by the Conservationists

The conservationists made their case stronger by offering concrete proposals. In a book entitled Goluboi serdtse Sibiri (Blue heart of Siberia) scientists proposed a general schema for the complex use of the natural resources of Lake Baikal.[74] Furthermore, a very distinguished group of Soviet professionals in May 1960 publicly requested that the following three measures be taken:
1. that approval for the scientifically unfounded Giprobum Organization projects be withdrawn, that work on the Baikal'sk and Selenga cellulose mills be stopped and the structures be dismantled;
2. that, in the economic and scientific interests of the country and for the good of the generations to come, Lake Baikal and its basin be declared a territory of extraordinary value to the country, so as to guarantee the uninterrupted natural replenishment of its waters and of the forests in its basin;
3. that this territory of extraordinary value to the country be administered by one agency responsible for the comprehensive utilization of its natural resources.[75]
Also in the mid-1960s the scientific institutions of the Siberian Division of the USSR Academy of Sciences, the USSR Institute of Geography, in collaboration with the State Planning Committee's Council for City Planning Institute, worked out a plan entitled "Basic Trends for a General Scheme for the Integrated Utilization of the Natural Resources of Lake Baikal and its Basin." This project appears to have been an official attempt at compromise because it

provides for the development in every possible way of the Baikal region's productive forces and at the same time ensures the full and guaranteed preservation of the lake's matchless scenery and the physical and chemical properties of its water.[76]

The project plans included provisions for the creation of the USSR's first national park. Two different versions were drawn up within which the proposed park's size ranged from 13,000 km^2 to 40,000 km^2. By comparison the largest park in the United States, Yellowstone, is less than 10,000 km^2. As suggested in the above quote, the park was envisaged as a comprehensive administrative institution that would exercise sanitary and technical control over the water transport, lumber, fishing, and chemical enterprises in the area. Facilities capable of accommodating 500,000 tourists annually were to be built in the following areas: Khamar-Daban and Barguzin Range areas, Cape Sviatoi Nos, Olkhon Island, Chivrokuiskii Bay, and along parts of the western shore from Peschanyi Bay to Bol'shoi Onguren.[77]

ADDITIONAL RESOURCE MANAGEMENT PROBLEMS OF THE LAKE BAIKAL BASIN

With this extensive background to the Baikal controversy it is now time to discuss the events that have actually transpired in the Baikal Basin from the mid-1960s until the beginning of 1973 in order to better evaluate the nature of the water quality problems of the area.

The preceding discussion of the Baikal controversy is somewhat misleading in that it tends to limit the potential and actual pollution threats to Baikal to essentially three sources: the Baikal'sk mill, the Selenginsk mill, and the indirect consequences of various timber procurement and transport techniques. In fact, the major thrust of the conservationist forces has been concentrated upon these three sources. However, other significant pollution sources exist.

Additional Sources of Water Pollution

The Selenga River and its tributaries, the Uda, Chikoi, Khilok, and Dzhida rivers (see Map 6.1) serve as receptacles for the industrial sewage from over 50 enterprises in Buriatiia. As of December 1968, only 18 of these enterprises possessed purification installations, and most of these facilities were unfinished.[78] From 1955 to 1965 the quantity of effluent discharged into the waterways of the Lake Baikal watershed increased by a factor of 24. During the same period the number of purifying installations at industrial enterprises only grew by a factor of 8.2, and because of overloading, the purification level reached was not satisfactory.[79] In 1968 a total of 62 enterprises were directly discharging a combined total of 123,000 m^3 per day of sewage effluent into the rivers and lakes of the Buriat Autonomous Republic. Of this total the Selenga was serving as the receptacle for 10,900 m^3 per day of effluent from 18 industrial enterprises and 36,000 m^3 per day of municipal sewage. The Uda was receiving 45,700 m^3 per day from 20 enterprises. Finally, 30,600 m^3 per day from 24 enterprises was flowing into other rivers and lakes of the republic.[80] Since this republic is largely within the Baikal watershed, the vast majority of this effluent load eventually has been emptying into the lake. This situation does not seem to have been improved significantly since 1968, although a recent report claims pollution of the Selenga River is being reduced.[81]

On the other hand, the mean discharge of the Selenga at Ulan-Ude is 79.9 million m^3 per day.[82] Thus, this volume would be sufficient to dilute all the daily wastes of the Buriat Autonomous Republic by a factor of 640. This suggests that the pollution problems along the Selenga are quite minor, except perhaps for restricted or limited areas near sewage outfalls.

Important individual point sources of pollution include the leather tanning plant located at Chikoi and the meat combine located at Ulan-Ude (see Map 6.1). The latter is the largest meat packing plant in Eastern Siberia. Micklin includes the sawmills at Il'inka and Babushkin as pollution sources (see Map 6.1).[83] If they are, the most probable waste would simply be wood, which, as mentioned earlier, adds to BOD levels. The oil depot at the mouth of the Barguzin River quite possibly may be a source of occasional oil spills and leakage.[84] Most of the other polluting enterprises are located in or near the capital city of the Buriat Republic, Ulan-Ude. The capital, with a population of over 200,000, is a major source of pollution.[85] As previously mentioned, the city was still, in late 1968, discharging all of its municipal sewage (36,000 m^3 per day) directly into the Selenga River with little prior purification treatment.[86]

The Series of Baikal Decrees

In 1966 the Russian Republic government was ordered to take a number of measures to prevent pollution of the Selenga and its tributaries. Accordingly, the Russian Republic Ministry of the Communal Economy was scheduled to issue design assignments in 1967 for the construction of a second section of Ulan-Ude's municipal sewerage system and for purification facilities. In 1969 B. Semenov, chairman of Buriat Autonomous Republic People's Control Committee, complained of lengthy delays and slow elaboration of these designs.[87]

The deadline for completion of construction of these facilities was scheduled for 1973.[88]

A Council of Ministers of the USSR decree made public in February 1969 again ordered the Russian Republic Council of Ministers and the various ministries and departments having enterprises in the Baikal Basin to ensure construction of appropriate purification and water-conservation installations in order to halt and prevent the discharge of unpurified communal, household, and industrial sewage into Lake Baikal and its tributaries.[89] A follow-up resolution announced in September 1971 ordered the Council of Ministers of RSFSR and the ministries having jurisdiction over the enterprises discharging waste water into the Selenga and its tributaries to inspect the state of construction of purification installations at the enterprises and to establish definite construction deadlines to expire not later than 1972.[90]

Thus, not less than six years will have elapsed between the initial order to take corrective and preventive measures and its actual implementation, and Semenov's 1968 criticism of some of the ministries and departments for not responding responsibly to the original 1966 order seems to have been justified. It is not insignificant that the latest resolution was signed by both the Communist party of the Soviet Union and the USSR Council of Ministers. The party's endorsement probably testifies to the earnestness of the Soviet effort to protect the Baikal Basin from pollution. At the same time, it suggests the likelihood that implementation of the original 1966 order has been quite slovenly. But in fairness to the Soviets, if this latest decree is in fact implemented in its entirety, the water pollution threat to Lake Baikal will be sharply curtailed.

The Decline of the Baikal Fishery

Although not emphasized previously, the conservationist forces fear that pollution has endangered the Baikal fishery. Hence, it seems appropriate to investigate this facet of the Baikal controversy. Table 6.1 lists the Lake Baikal commercial catch of omul (whitefish), the major commercial fish in Lake Baikal, for selected years from 1942 to 1967 and shows a continuous decline in the catch, at least since the mid-1950s.[91] In fact, it became so acute that commercial fishing of omul in Lake Baikal was eventually banned for a period of six years beginning in 1969.[92] An obvious question that comes to mind is whether or not pollution is responsible for this decline in fishing.

The answer appears to be a qualified yes because pollutants and sunken timber from river rafting along the Selenga have destroyed many spawning areas. However, as of yet the Baikal'sk and Selenginsk mills are at most insignificant contributing factors.[93]

TABLE 6.1

Annual Catch of Lake Baikal Omul for Selected Years,
1942-67

Year	Catch in Centners (100s kg)
1942	91,300[a]
1945	91,300[b]
1949	41,800[a]
1951-55 (average)	63,000[e]
1957	40,197[b]
1959	31,000[a]
1963	33,200[c]
1964	26,500[c]
1965	20,900[c]
1966	18,000[c] [d]
1967 12,000 (plan)[d]	10,000 (actual)[c] [e]

Sources: (a) V. Baraev, "The Baikal Whitefish . . . ," Current Digest of the Soviet Press 12 (November 23, 1960): 24; (b) Marshall Goldman, The Spoils of Progress (Cambridge, Mass.: MIT Press, 1972), p. 182. Original source: B. R. Buiantuev, K narodnokhoziaistvennym problemam Baikala [In regard to the economic problems of Baikal] (Ulan-Ude: Buriatskoe knizhnoe izdatel'stvo, 1960), p. 22; (c) G. I. Galazii, Baikal i problema chistoi vody v Sibirii (Irkutsk: Academiia Nauk SSSR, 1968), p. 23; (d) L. Gribanov, "Time Will Tell," Current Digest of the Soviet Press 19 (December 20, 1967): 8; (e) B. Moskalenko, "Once More About the Fate of the Lake Baikal Omul," Current Digest of the Soviet Press 19 (December 20, 1967): 9.

Short-Sighted Fishing Practices

Other nonpollution or nonforestry factors seem to have been major causes in the deterioration of the omul catch. These include poaching, irrational commercial fishing practices, and slow progress in the creation of fish hatcheries. Basically poaching is of two types: poachers who catch the fish illegally for private profit, and "legal" poachers who catch undersize fish "for the sake of the plan." In 1959 the East Siberian State Hatchery Trust recorded 23 cases of fishing artel* brigades being apprehended for catching undersize fish. The

*A cooperative association.

same year in Buriatiia alone 514 cases of violations of fishing regulations were recorded, 23.5 tons of omul and 19,100 meters of net were confiscated. Only 128 cases went to court and the prosecutor's office dismissed more than half "because the crimes were too petty."[94]

Representatives of the fishing industry have even advocated the catching of undersize fish, which is the major offense of the "legal" poachers. In June 1966 Comrade Gribanov, Russian Republic deputy minister of the Fishing Industry, required from the USSR Ministry of the Fishing Industry: (1) legalization of the large-scale catch of undersize omul, (2) permission to catch fish weakened by spawning, and (3) permission to change other aspects of the fishing regulations.[95]

Similarly, in 1967 the director of the East Siberian Fishing Industry Administration secured permission to catch undersize omul and to postpone the ban on fishing in the prespawning period. Furthermore, in the summer of 1967, Comrade Dossov, assistant director of the Chief Fish Hatcheries Administration, signed and sent to the All-Union Fishing Ministry recommendations for the reduction in size of the mesh of omul net and for the complete abolition of industrial minimum sizes of omul.[96] The desire to fulfill production plans probably motivated these various shortsighted measures.

Delays in Construction of Fish Hatcheries

In 1960 the Buriat Economic Council decided to build three artificial breeding plants for whitefish and salmon on the Upper Angara, Selenga, and Greater Chivyrkui rivers.[97] From 1966 to 1968 only 30 percent of the appropriated funds had been spent and by 1968 only the Chivyrkui plant was under construction.[98] It appears that this hatchery plus another one, Bol'sherechenskii, located near the Selenga delta, are now complete.[99] Because of faulty planning, the fishing industry must share part of the responsibility for the decline of the Baikal fishery.

The Damage to the Fishery from Pollution and Logging

Pollution and logging have also caused serious losses of whitefish through the drastic deterioration and in some cases complete destruction of its native spawning grounds. Work of the Siberian Fishing Industry Research Institute shows that pollution of the Selenga— which is the major omul spawning river—has caused the massive destruction of omul roe at the spawning beds. According to data of the Limnological Institute the quantity of spawning omul killed over the past five years has increased four times.[100] L. Gribanov, who is criticized earlier in the text, claims that "concentrations of certain substances discharged into the Baikal Basin exceed the permissible norms by hundreds and thousands of times."[101] Unfortunately Gribanov's claim, which may well be valid, loses credibility because of his attempt to introduce irrational fishing practices. Perhaps he is trying to shift total blame for the declining fishery upon the timber and other industries of the area.

Furthermore, many river mouths of great importance for spawning are cluttered with timber and river beds are lined with sunken logs. The Turka, Barguzin, and Upper Angara, in addition to the Selenga, have been major spawning areas. Unfortunately the Barguzin and Turka have lost spawning significance because of timber rafting.[102] The Barguzin, Turka, Selenga, and Goloustnaia (see Map 6.1) all have timber rafting problems.[103] Particularly unfavorable conditions for spawning resulting from timber floating exist along some 25 km of the Selenga upstream from Il'inka to Tataurova.[104] In fact, from 1958 to 1968 an estimated 1.5 million m^3 of timber sunk. As a result, 50 streams totaling over 3,500 km have been lost as spawning areas for fish.[105] In addition to the destruction of fish roe and fish spawning beds, logging and discharges of municipal and industrial effluents have resulted in a measurable change in the chemical composition of the water of the Selenga and Barguzin rivers.[106] As a result, the water of these rivers has a higher concentration of mineral salts than the water of Lake Baikal.[107] However, considering the high river discharge to effluent ratio previously stated, it seems more plausible that the direct and indirect effects of increased logging are more important than pollution in accounting for the observed hydrochemical changes of the rivers' regimes.

The Poor Execution of Promises by the
Timber Interests

One of the assurances given by the timber interests had been the cessation of rafting and the introduction of railroad transport.[108] Accordingly, it seems prudent now to judge the validity of the conservationists' fears associated with intensified logging. In January 1966, less than a year before the Baikal'sk mill began production, Oleg Volkov complained that the timber procurement agencies had made no preparations for selective cutting, that is, timber transport roads were not being built and special technology such as aerial cableways was not being provided. Equally important, he claimed the 1960 decree of the RSFSR Council of Ministers, which banned tree cutting in about 1.5 million hectares of forest and created water conservation zones, was not being properly implemented. To support this claim he cited the example of the Sliudionka lumber camp where in 1966 some 150,000 m^3 of timber was clear-cut and this in an area where only thinning and hygienic and restoration cutting was supposedly allowed.109

Taking the preceding items into account, it is not surprising that in December 1967, slightly more than a year after the Baikal'sk plant began operation, floatation was still practiced; although a railroad had been built, freight cars had not been supplied.110 Apparently, for the timber executives, production at the plant was more important than the damage inflicted upon the rivers and fish by continued rafting. One need not belabor examples of parallel executive behavior in the United States.

UNFULFILLED PROMISES AND THE BAIKAL DECREES

Probably in response to these unfulfilled assurances upon the part of the timber and pulp interests, the previously mentioned decree of February 1969 called for a number of timber conservation measures:
1. A water-conservation zone was to be established for Lake Baikal within the boundaries of its watershed.
2. Appropriate organizations were charged with the termination of lumbering operations in all state forests in the Lake Baikal Basin.
3. Logging was to be prohibited on slopes over 25° that perform water and soil conservation roles, in steppe and wooded-steppe pine forests, in forests of recreational value, and in nature preserves.
4. River channels of the basin were to be dredged to remove sunken and decaying logging debris.
5. Cableways were to replace log skidding to reduce erosion.

The USSR government charged the USSR People's Control Committee with control for the implementation by the ministries and departments of the above measures.111

On September 24, 1971, a third decree on Baikal was issued under the title "On Additional Measures to Ensure the Rational Utilization and Conservation of the Natural Resources of the Lake Baikal Basin."11 In addition to instructing the ministries and departments involved in the use of Lake Baikal's resources and the USSR Academy of Sciences to speed up the working out of plans for the water conservation zones stipulated in the 1969 decree, the party Central Committee and USSR Council of Ministers issued several instructions to the USSR Ministry of the Lumber and Wood-Processing Industry:
1. to cease completely, no later than 1973, the loose floating of timber along Lake Baikal tributaries;

2. to carry out, no later than 1972, the transport of timber on the lake in wave-resisting rafts (sigary);
3. to ensure during 1971-75 the cleaning of river beds of sunken timber; especially the spawning grounds of the Itantsa, Angyr, and Barguzin and their tributaries, the mouths of the rivers entering Barguzin Gulf, and the shallow areas of the Selenga;
4. to elaborate a 10-year plan for forest felling in the basin;
5. and to procure wood in the basin in strict conformity with the established regulations of the area.[113]

THE PRESENT STATUS OF LOGGING IN THE BAIKAL BASIN

Literaturnaia gazeta on April 5, 1972, gave a progress report on the implementation of these and other measures.[114] For example, free floating of timber upon the Barguzin River was expected to be ended in 1972. However, this objective, as well as the cessation of timber rafting along the Uda River, apparently was not accomplished in 1972.[115] In 1970 Buriatiia received over 100 Japanese log trans-porters that are presently at work. A logging road is being finished along the Turka River—50 km are complete and 30 km remain to be finished. Machines are actively removing timber in the Turka estuary and assembling them into sigary for dispatch to the port of Vydrino near Baikal'sk. Sergei Grishkin, director of the Turka lespromkhoz (forestry enterprise) claims the forest reorganization as prescribed in the 1971 decree has been executed. Forest categories have been changed and water conservation zones created. A long-term forest felling program has also been elaborated incorporating strict rules for selective cutting. New logging roads totaling 600 km have been built. The cleaning of the rivers has begun with the Uda, Kurba, Dudun, Itantsa, Ona, and Turka rivers already cleared of discarded and submerged logs.[116]

The northernmost timber enterprise, the Kurumkansk lespromkhoz expects 1972 to be the final year of log floating with 220,000 m³ planned to be floated down the Barguzin River. A logging road 214 kilometers long from Ust'-Barguzin to the village of Maiskii has been completed, but only in the winter version—bridges are not finished. Six million rubles will be necessary to complete the road, whereas the annual volume of work of the contractor, the Barguzin Construction Materials Department, is only 1.5 million rubles. Also, the construction trust renovating the port of Vydrino for the introduction of rafts is behind schedule.[117] New rules for cutting trees, new felling quotas, and new forest classifications are reported to have been made.[118] Hence, although significant progress has been made, still much remains to be done in order for the forestry sections of the 1971 decree to be fully implemented on schedule.

THE CURRENT STATUS OF THE SELENGINSK CELLULOSE-CARDBOARD ENTERPRISE

Before reaching any tentative conclusions about the Baikal controversy the events at the Selenga and Baikal'sk mills must be brought up to date. The conflict between officials of the Ministry of the Lumber and Wood-Processing Industry and Ministry of the Fishing Industry over the type of purification facilities to be constructed at the Selenginsk Cellulose and Cardboard Combine has hindered the construction program. By December 1967 only 4 percent of the budget earmarked for purification of sewage had been spent and the 1968 assignment was not expected to be fulfilled.[119] On the other hand, during the first four or five years of the plant's operation, the sewage was supposed to be shunted into an immense receptacle being built at Kliukvennyi Falls, a huge depression 13 kilometers from Selenginsk. Furthermore, over 70 percent of the water was to be recycled.[120] The 1969 decree instructed the Ministry of the Pulp and Paper Industry to ensure that purification facilities were constructed at the Selenga Combine before production was commissioned.[121] The 1971 party-government resolution instructed the USSR Ministries of the Pulp and Paper Industry and Industrial Construction to finish construction of the first section of the Selenga factory in 1972. At the same time, it reiterated the 1969 decree by prohibiting the Ministry of Pulp and Paper Industry from commissioning the factory before completion of the purification complex.[122]

The director of the Selenginsk combine, Anatolii Vasilievich Goncharov, has said, "I on no account will commission the combine into a bit of functioning without the purification installations. . . . The Combine will enter the industrial world dressed spick and span and clean as in a parade."[123] The purification installations being built provide for mechanical, biological, chemical, and aeration treatments, with an hourly control to be kept over the effluent. By-products are to be used and recirculation will decrease the water demands of the combine by a factor of six. According to the chief plant technologist, Mikhail Borisovich Gainutdinov, during spawning the omul will not come into any contact with even purified effluent. At this time and during other emergencies, the effluent will be directed as originally planned into Kliukvennyi Falls.[124] While this appears to be a workable solution, it has at least one potential drawback. If emergencies occurred during the winter, it seems plausible that freezing would interfere with the shunted effluent flow.

In spite of the rather favorable report, much work remains unfinished on the purification facilities. While over half of the 1972
budget was earmarked for pollution control,[125] A. V. Georgiev, chairman of the Council of the Union's Conservation Committee, reported
in the September 21, 1972, edition of Izvestiia that both the construction of the purification installations and the utilization of capital investment funds allocated for them were still unsatisfactory at the
Selenga site.[126] Unfortunately, if the Baikal'sk experience serves
as any example, the good intentions of the director may not be enough
to prevent commissioning before the facilities are complete.

BAIKAL'SK 1966-70: A HISTORY OF FAILURES

The Baikal'sk mill was scheduled to go into operation in 1963
and to begin full capacity operation in 1965.[127] After numerous delays
partially associated with the conflict between the developers and conservationists, the Baikal'sk cellulose plant was put into operation in
1966 with unfinished purification facilities, in violation of existing
instructions.[128]
Why did this happen? Could not the Ministry of the Fishing Industry have prevented the commissioning? N. P. Alexeeva, chief of
the Department for the Conservation of Fish Reserves and the Regulation of Fishing of the State Production Committee for the Fishing Industry, said, "No, we do not have the right to stop the operation of the
enterprise. Only the Sanitary Inspection Service is permitted to do
that."[129] Iu. E. Donitov, USSR Chief Sanitary Physician and Deputy
Minister of Public Health stated, "We investigate only the sanitary
condition of waterways and may forbid the discharge of sewage water
in a river only when the contamination of nearby drinking-water supply exceeds health standards."[130]
Furthermore, P. Shternov, a member of the collegium of the
USSR Ministry of Land Reclamation and Water Resources, disputed
the prudence of achieving "complete independence" of the production
and purification phases of the Baikal'sk Pulp Plant operation by vest-
ing responsibility over the purification phase with his ministry rather
than with the plant executives as was and is the case at present.[131]
The above items are clear examples of bureaucratic buck-passing.
Although the purification installations were expected to cost 10
percent of the total construction budget,[132] there is evidence that the
State Committee for the Lumber, Pulp and Paper, and Wood-Processing Industry, and Forestry tried to invest as few funds as possible
in the construction of wood-chemistry enterprises in order to obtain
good production indexes per ruble of capital investment. The obvious
area for short-cuts was in pollution abatement.[133] This reasoning

may account for the fact that the sewage line to the Irkut River has never been built.[134] However, as Micklin states, the severe freezing climate would probably create technical problems difficult to resolve.[13] Volkov, in February 1966, commented on another technical problem at the plant; at that time the biological stage of purification still had no activated sludge and would not be able to prepare any until spring.[136]

By October 1967, after seeing the enormous purification complex and its "emergency capacity" at Baikal'sk, the staunch and long-time critic, Oleg Volkov, seemed convinced that the Baikal'sk mill would achieve the design indexes for sewage purification. He was impressed by such facts as the timber industry's quick decision and a 2,500,000-ruble appropriation to construct additional sedimentation tanks that will become necessary once the mill begins full production. The purified effluent was stated as being slightly yellowish and containing a higher mineral concentration than Baikal's water, but it was drinkable. Volkov thought that violations of the norm "should in fairness be attributed to the inevitable expanses of the start-up and adjustment period."[137]

Nevertheless, he still made some criticisms and cautionary statements. First, he mentioned the dilemma of emergency situations when the choice becomes either to disrupt the production plan or to pollute the lake. He concluded that unfortunately "the plant management is held responsible first and foremost for basic production."[138] For this reason Volkov suggested separate managerial responsibility for the purification and production phases of the pulp mill. As mentioned previously, P. Shternov, an executive of the USSR Ministry of Land Reclamation and Water Resources, argued against this idea. Furthermore, this suggestion has not been acted upon.

Second, Volkov lamented the high labor turnover at the Baikal'sk plant purification installations. The low level of amenities at Baikal'sk as well as the considerably higher salaries offered elsewhere, such as at Bratsk, were probably responsible for the departure of 100 of the purification facilities' 230-240 employees during the summer of 1967.[139]

Third, interruptions in the timber supply were threatening the biological stage of treatment. Any long production stoppage can destroy the bacteria of the activated sludge.[140]

In a direct response to Volkov's newly found optimism, G. Gorin, a Candidate of Technical Science, cautioned that the plant would discharge over 60 million m^3 per year of the "yellowish" effluent into the lake and that the purification process would intercept only 97 percent of the impurities.[141] However, at the present state of pollution control technology, 97 percent is very respectable anywhere in the world. Furthermore, 60 million m^3 is insignificant compared to Baikal's volume of 23,600 km^3 (1 km^3 = 1 x 10^9 m^3).[142]

In January 1968 the USSR Minister of the Fishing Industry, A. Ishkov, also wrote a retort to Volkov's optimism. He cited 14 serious flaws in the purification plant's operation. Although the purified effluent was claimed to be largely within the design limits with only periodical violations of the permissible norms, this situation was being achieved by a substantial dilution of the effluent with pure Baikal water. Other objections were that (1) the so-called black sewage regularly exceeded the design norms for content of suspended solids, mineral salts, BOD, and other indexes; (2) the white sewage containing a low concentration of organic solids was being discharged directly into the lake, by-passing biological treatment; (3) the technological regime for the use of Vako filters was not being maintained; (4) there had been instances of ash-dump drainage into the lake; and (5) Baikal'sk was not purifying domestic sewage.[143]

The following are some examples of recent pollution control problems at the Baikal'sk cellulose plant. According to data of the Hydrometeorological Service, 383 tons of toxic substances were discharged into the lake from mid-1967 to mid-1968. The research vessel Vereshchagin of the Limnological Institute detected foul-smelling sulfur compounds (sulfides, mercaptans) at distances up to 20 kilometers from Baikal'sk and at depths up to 700 meters.[144] Zones of pollution ranging up to 10-13 km^2 and lasting for up to two months were observed periodically during the mid-1960s.[145] Moreover, on July 1, 1968, a zone of pollution from the Baikal'sk cellulose plant 30 km long and 5 km wide was observed along the south shore of the lake.[146] In early 1969 the Limnological Institute reported that the flora and fauna near the sewage outfall had diminished by one-third to one-half.[147]

Even Trofimuk and Gerasimov's hypothesis that the poor design of the intake and outfall pipes would result in self-pollution seems to have come true. Accordingly, the plant has had to provide supplementary treatment, and by 1970 the planned 20 million ruble total investment in pollution control had in fact exceeded 40 million rubles. Furthermore, the planned treatment cost of 4 kopecks per m^3 had climbed to 10 kopecks per m^3, or approximately 20,000 rubles per day.[148]

These large cost overruns lend themselves to two contrary interpretations. They may just be examples of deliberate underestimation of costs on the part of the planners and timber industry executives in order to win initial approval for the plant's construction, or they may signify at least a partial victory for the conservationist forces, that is, substantially more resources have been committed to the task of curbing pollution.

Galazii claimed that the effluent norms listed in Table 6.2 were violated over 100 times in 1968 and that from January to May 1967 there was not a single ten-day period without violations of the planned

TABLE 6.2

Permissible Concentrations of Individual Chemical
Components in the Sewage (waste-water effluent) of
Various Industrial Production for Lake Baikal

Component	Maximum (limit) Concentration (milligrams per lit
Carbonate (HCO_3)	76.0
Sulfate (SO_4)	6.0
Chloride (Cl)	1.0
Floride (F)	0.3
Bromide (Br)	0.003
Iodide (I)	0.001
Nitrate (NO_3)	0.5
Nitrite (NO_2)	0.00
Phosphate (PO_4)	0.1
Calcium (Ca)	20.0
Magnesium (Mg)	4.0
Sodium (Na)	5.0
Potassium (K)	3.0
Ammonium (NH_4)	0.05
Iron (Fe-total)	0.06
Aluminum (Al)	0.05
Chromium (Cr)	0.006
Manganese (Mn)	0.005
Copper (Cu)	0.006
Nickel (Ni)	0.005
Cobalt (Co)	0.002
Silicon dioxide (SiO_2)	5.0
Lead in nonorganic compounds (Pb)	0.05
Arsenic (As)	0.01
Zinc (Zn)	0.01
Cadmium (Cd)	0.005
Ammonia (free) (NH_3)	0.05
Chlorine (free) (Cl)	not permitted
Sulfides	not permitted
Tannic acid	less than 1.0
Resinous substances, washed out of wood	1.0
Oil and petroleum products in dissolved and emulsified state	0.05
Organic substance	
a) oxidation by permanganate	5.0
b) BOD_5	0.8-1.0
c) COD	6.0
d) other organic substances not naturally characteristic of waterways	not permitted
Oxygen (O_2)	not less than 8.
Suspended substances	0.2-0.3

Source: G. I. Galazii, Baikal i problema chistoi vody v Sibiri (Irkutsk: Academiia Nauk SSSR, 1968), pp. 49-50.

norms.[149] Furthermore, during a visit to the pulp mill Galazii noted that several sections of the treatment complex were inoperative.[150] Even more alarming are the data of the Baikal Basin Inspectorate. According to these data about 500 instances of violations of effluent norms occurred in 1967, 540 occurred in 1968, and in 1969, despite the enactment by the USSR Council of Ministers of the Baikal conservation decree, 1,186 cases were recorded. Thus, the pollution control problems actually intensified rather than diminished after the signing of the law,[151] since according to the 1969 decree the Baikal'sk mill was ordered to complete construction of its waste facilities and shops for the utilization of industrial by-products.[152]

More importantly, through various forms of subterfuge and despite the protests of the Baikal Basin Inspectorate of the Ministry of Land Reclamation and Water Management, in 1969 the USSR Ministry of the Pulp and Paper Industry developed new norms that drastically reduced the specifications for the quality of the purified effluent.[153]

THE STATUS OF BAIKAL'SK TODAY

The joint signing on September 24, 1971, by the Central Committee of the Communist party of the Soviet Union and the Council of Ministers of the USSR of the Law on the Preservation of Lake Baikal[154] testifies both to a growing Soviet commitment to protect the lake and to the slow implementation of a similar February 1969 decree issued by the Council of Ministers of the USSR.[155] Some sections of this latest law are directly addressed to the Baikal'sk pulp mill. Accordingly, these sections call upon the Ministry of the Pulp and Paper Industry (1) to complete the sewage treatment facilities in 1971; (2) to ensure commissioning in 1971 of the second section of the pulp plant and shops for the utilization of production wastes; and (3) to complete the entire Baikal'sk complex in 1973.[156]

What is the situation at Baikal'sk today? Literaturnaia gazeta reports that the plant has not violated the norm for some months. But this really is not welcome news. On the contrary, the norms at the end of 1971 had again been relaxed so much that it would be difficult for the effluent to exceed them. The director of the mill, Aleksandr M. Senchenko, is pleased with the new norms. "We are within the norm. Baikal is a powerful aerator. Don't become worried. Its ability for self-purification is colossal."[157] Both the signatures of Deputy Minister A. Burnazian of USSR Ministry of Public Health and Deputy Minister S. Studentsk of the USSR Ministry of the Fishing Industry sanction these new, but relaxed, norms.[158]

Burnazian justifies his signature by claiming that after dilution with Baikal water, the water will not be harmful to either man or animals. Studentsk claims the papermakers requested the new norm as a temporary measure for two years while the plant is undergoing its trial period. For perspective, the plant has already been in trial operation about five and one-half years! Because the many fishing areas are quite distant from Baikal'sk, the new norm will probably not harm the fishing industry, at least in the short run. Evgenii K. Fedorov, manager of the Chief Administration of the Hydrometeorological Service of the USSR, opposes the new norms and is closely monitoring the conditions of the Baikal'sk plant's effluent. The USSR Ministry of Reclamation and Water Management also objected to the new norm, but the papermakers got their way.[159] It will be seen that to this day the controversy over the Baikal'sk Pulp Mill has not been thoroughly resolved and the timber executives continue to practice subterfuge and evasion in order to avoid compliance with control instructions. Even the latest report is not overwhelmingly optimistic about the prospects of any significant improvement in the near future.

TENTATIVE CONCLUSIONS

As stated at the beginning of the chapter, this author is in substantial agreement with Goldman's thesis that the Soviet Union's ideological disdain for placing monetary value upon natural resources and its preoccupation with industrial production output or "the plan" leads to "environmental disruption."[160] In other words, by treating natural resources as free commodities, the Soviet planners and decision maker. experience great difficulties in trying to internalize social costs into the planning and day-to-day operations of their economy. Unfortunately the U.S. experience is quite similar. Until recently we, too, have placed a very low value upon air and water. As in the United States, bureaucratic in-fighting, buck-passing, subterfuge, and evasion are practiced by Soviet governmental officials, planners, and industrial executives. The Baikal case study as presented in this chapter testifies to the existence of all of these resource management problems, difficulties, and conflicts.

It will take perhaps until 1980 to really test the determination and ability of the Soviets to halt the present trend and to prevent future pollution of the waters of the Lake Baikal Basin. And it may take longer before the full impact on the lake is known. Presently a water pollution threat, of course, still exists.

But the magnitude or seriousness of this threat is open to different interpretations. Goldman claims "however honorable his intentions, man's efforts to 'harness' Lake Baikal have been catastrophic, and concludes that "Lake Baikal is too fragile to withstand even the mildest maltreatment."[161] These assertions may be overstated.

First, industrial development within the Baikal Basin and particularly along the lake's shoreline is both recent and so far minor in comparison with many of the waterways of European Russia. The area is still sparsely settled with no sizable urban settlements along its shore and only one medium-sized city, Ulan-Ude (200,000 pop.), within its watershed. Surely this situation is more analogous to Lake Tahoe in California than it is to Lake Erie. Goldman has recently acknowledged that Baikal is definitely not another Lake Erie in terms of pollution.[162] In fact, in terms of its remoteness from the mass of the Soviet citizenry, Baikal is similar to the remoteness of Canada's Great Slave Lake or Great Bear Lake from the citizenry of Canada and the United States.

Second, the enormous volume of water contained in the lake plus its extremely high concentration of dissolved oxygen contribute strongly to the tendency for environmental stability rather than environmental fragility.

Finally, even with the latest relaxation of the effluent norms, the Baikal'sk mill's purified effluent is still drinkable. Goldman also acknowledges this situation.[163] Furthermore, Professor Galazii, the long-time critic and pollution pessimist of the Limnological Institute has recently been quoted as saying, "I think we nipped it in the bud."[164] However, in fairness, this statement was made a short time before the latest relaxation of the norms.

Perhaps the most interesting and significant fact about the Baikal controversy is simply that the debate occurred at all, let alone on such a scale that even a major Soviet movie has been made about it.[165] Entitled "At the Lake," it has been awarded the coveted State Prize.[166] Farley Mowat claims the planners originally planned five plants, and that the conservationist forces scored a victory by forcing the cancellation of three of them.[167] In addition, he quotes a Soviet zoologist as optimistically claiming:

> We have done more than save Baikal. The fight woke up
> the whole of the Soviet Union to one of the grave dangers
> threatening mankind. We will not go to sleep again. Our
> leaders now understand how great the danger is and they
> are really listening hard to those who can tell them how
> to control and stop the damage done by a thoughtless
> modern industrial society. What happened here at Baikal
> will help set the pattern for the future development of
> our country.[168]

Hopefully this optimism will prove to be warranted. Regardless, it does testify to the rather profound scope and scale with which the Baikal conservation struggle has been waged in the Soviet Union. On the other hand, the conservationists have failed to take cognizance of the Baikal'sk mill's air pollution problem.[169]

It is the "intuitive judgment" of this author that the pulp plants should never have been built. This "judgment" is based simply upon the desire to preserve the physical and biological uniqueness and beauty of Baikal. However, such sentiments come easy when one is thousands of miles away. Plausibly the workers of Buriatiia, 30 percent of whom are directly employed by the timber industry, might well intuitively feel quite the opposite.[170]

To sum up, the overriding fact is that the plants do exist. Moreover, a recent article by I. Novikov, vice-chairman of the USSR Council of Ministers and chairman of the USSR State Construction Committee, indicates that more rather than less economic development is in store for the Lake Baikal Basin.[171] On the other hand, the Baikal Zapovednik (nature preserve) has been created,[172] and 60 new positions of the USSR People's Control Committee have been created to oversee the protection of Lake Baikal and its natural resources.[173] Nonetheless, the fundamental question that at present still remains unanswered is whether genuinely adequate environmental safeguards will accompany the future economic development in the Baikal Basin as well as elsewhere in the USSR. Past experience—particularly at Baikal'sk—weighs heavily against this course of action. However, the party's signature on the 1971 Baikal decree,[174] the recent Soviet-American Environmental Protection Agreement,[175] and the U.S.-USSR memorandum that establishes Lake Baikal as a water pollution study area,[176] bring hope that the Baikal controversy will be remembered best not as another example of Soviet man's wanton exploitation of the natural environment, but as a significant turning point in Soviet environmental consciousness.

NOTES

1. Philip P. Micklin, "The Baykal Controversy: A Resource Use Conflict in the U.S.S.R.," Natural Resources Journal 7 (October 1967): 485-98; Theodore Shabad, "Soviet Moves to Halt the Pollution of Lake Baikal," New York Times, February 8, 1969, sec. 1, p. 8, and "News Notes," Soviet Geography: Review & Translation 10 (March 1969): 145; Philip R. Pryde, "Water Pollution in the Soviet Union," Proceedings of the Association of American Geographers 2 (1970): 110-15, and Conservation in the Soviet Union (London: Cambridge University Press, 1972), pp. 147-51; Marshall Goldman, "The Convergence of Environmental Disruption," Science 170 (October 2, 1970):

37-42, "Our Far-Flung Correspondents: The Pollution of Lake Baikal," The New Yorker, June 19, 1971, pp. 58-66, and Marshall Goldman, The Spoils of Progress: Environmental Pollution in the Soviet Union (Cambridge, Mass.: M.I.T. Press, 1972), pp. 177-209; Victor L. Mote, "The Geography of Air Pollution in the U.S.S.R." (Ph.D. dissertation, University of Washington, 1971), pp. 61-62.

2. U.S. Department of State, Memorandum of Implementation of the Agreement between the United States of America and the Union of Soviet Socialist Republics on Cooperation in the Field of Environmental Protection of May 23, 1972, Press Report No. 236 (1972), p. 3.

3. A. V. Plashchev and V. A. Chekmarev, Gidrografiia SSSR [Hydrography of the USSR] (Leningrad: Gidrometeorologicheskoe izdatel'stvo, 1967), p. 258.

4. A. N. Baranov, ed., Atlas SSSR [Atlas of the USSR], 2d ed (Moscow: Glavnoe upravlenie geodezii i kartografii pri Sovete Ministrov SSSR, 1969), pp. 58-59, 70-71, 74-75; L. Rossolimo, Baikal (Moscow: Izdatel'stvo "Nauka," 1966), pp. 24-25.

5. Baranov, Atlas SSSR, p. 152; V. Sharov, "Baikal in Autumn: Man and Nature," Current Digest of the Soviet Press 20 (December 18, 1968): 28. Translated from Pravda, November 27, 1968, p. 6; K. O. Dobronravov, ed., Baikal (Moscow: Izdatel'stvo "Mysl'," 1971), introduction; A. A. Trofimuk and I. P. Gerasimov, "Sokhranit' chistotu vod ozera Baikal" [Preserve the purity of Lake Baikal's water], Priroda, no. 11 (November 1965), p. 50.

6. V. I. Astrakhantsev, "Zashchitit' vodnye resursy Irkutskoi oblasti ot zagriazneniia" [Protect the water resources of Irkutsk Oblast from pollution], Gidrotekhniki i melioratsiia 21 (January 1969): 102; Baranov, Atlas SSSR, p. 58; Plashchev and Chekmarev, Gidrografiia SSSR, p. 258; Dobronravov, Baikal, introduction.

7. Oleg Volkov, "The Call of Lake Baikal," Soviet Life, August 1966, p. 6; Dobronravov, Baikal, introduction; I. Novikov, "Bogatstva Baikala—strane" [The wealth of Baikal is the wealth of the country], Pravda, October 27, 1971, p. 3.

8. Author's calculation using 23,000 km3 for Baikal's volume and data on the world's water supply from Tinco E. A van Hylckama, "Water Resources," in Environment: Resources, Pollution and Society, William W. Murdoch, ed. (Stamford, Conn.: Sinauer Associates, Inc., 1972), p. 139.

9. Rossolimo, Baikal, p. 62; Baranov, Atlas SSSR, p. 58.

10. Dobronravov, Baikal, introduction; Plashchev and Chekmarev, Gidrografiia SSSR, pp. 236-38.

11. Baranov, Atlas SSSR, pp. 70-71, 86-87.

12. Plashchev and Chekmarev, Gidrografiia SSSR, p. 264.

13. Rossolimo, Baikal, pp. 91-94.

14   Trofimuk and Gerasimov, "Sokhranit' chistotu vod ozera Baikal," p. 52.

15. Rossolimo, Baikal, pp. 98-102.

16. Dobronravov, Baikal, introduction.

17. Goldman, The Spoils of Progress, p. 183. Original source: Komsomolskaia pravda, May 11, 1966, p. 2.

18. M. Markelov, "Vostochno-Sibirskii kompleks" [East Siberian complex], Izvestiia, July 31, 1963, p. 3.

19. "Posle togo kak vystupili 'Izvestiia': 'Tiazhba na Baikal'" [Follow-up: 'controversy at Baikal'], Izvestiia, August 1, 1962, p. 3.

20. A. Merkulov, "Alarm from Baikal," Current Digest of the Soviet Press 17 (March 24, 1965): 24-26. Translated from Pravda, February 28, 1965, p. 4.

21. Boris Konstantinov et al., "Baikal Waits," Soviet Life, August 1966, pp. 6-7. Translated from Komsomolskaia pravda, May 11, 1966, p. 2.

22. Iu. Danilov, "Let Us Protect the Water, Air, and Soil from Pollution," Current Digest of the Soviet Press 17 (July 7, 1965): 13-14. Translated from Pravda, June 21, 1965, p. 2.

23. M. A. Sholokhov, "Speech by Comrade M. A. Sholokhov, Writer, Rostov Party Organization," Current Digest of the Soviet Press 18 (May 11, 1966): 26-27. Translated from Pravda, April 2, 1966, p. 5.

24. Oleg Volkov, "A Writer's Notes: A Trip to Baikal," Current Digest of the Soviet Press 18 (February 23, 1966): 14-15. Translated from Literaturnaia gazeta, January 29, 1966, pp. 1-2.

25. A. Merkulov, "Stroika u Baikal" [Construction project at Lake Baikal], Pravda, September 21, 1960, p. 6.

26. "Na beregu Baikal" [On the shore of Lake Baikal], Izvestiia, June 22, 1960, p. 3.

27. Goldman, The Spoils of Progress, p. 183.

28. "Ozhila taezhnaia dolina" [A taiga valley has come alive], Izvestiia, August 24, 1960, p. 3.

29. Merkulov, "Stroika u Baikal," p. 6.

30. Goldman, The Spoils of Progress, p. 183.

31. Ibid., pp. 183-84. Original source: Komsomolskaia pravda, December 26, 1961, p. 4; G. I. Galazii, Baikal i problema chistoi vody v Sibiri [Baikal and the problem of pure water in Siberia] (Irkutsk: Academiia nauk SSSR, 1968), p. 12.

32. L. Shinkarev, "Tiazhba no Baikal" [Controversy at Baikal], Izvestiia, July 20, 1962, p, 4; "Posle togo kak vystupili 'Izvestiia': 'Tiazhba no Baikal'," p. 3.

33. For example, see Pravda, February 28, 1965; Literaturnaia gazeta, February 6, March 18, April 10, 13, 15, 1965; January 29, June 2, October 6, 1966; Izvestiia, September 25, November 18, 1965; Komsomolskaia pravda, May 11, 1966.

34. Micklin, "The Baykal Controversy," pp. 485-98; Goldman, "Our Far-Flung Correspondents," pp. 58-66, and The Spoils of Progress, pp. 177-209.

35. Shinkarev, "Tiazhba na Baikal," p. 4.

36. Oleg Volkov, "Priroda i my: Tuman nad Baikal" [Nature and we: fog over Baikal], Literaturnaia gazeta, February 6, 1965, p. 1, and "A Writer's Notes," p. 14.

37. Merkulov, "Alarm from Baikal," p. 25; A. Trofimuk, "Tsena vedomstvennogo upriamstva: otvet Ministru SSSR G. M. Orlovu" [The cost of bureaucratic stubborness: an answer to minister of the USSR, G. M. Orlov], Literaturnaia gazeta, April 15, 1965, p. 2.

38. Trofimuk and Gerasimov, "Sokhranit' chistotu vod ozera Baikal," pp. 53-54; Galazii, Baikal i problema chistoi vody v Sibiri, p. 9.

39. Merkulov, "Alarm from Baikal," p. 25.

40. Volkov, "The Call of Lake Baikal," p. 8.

41. Author's calculation: 23 x 10^{12} m^3 = volume of Lake Baikal, 1 x 10^6 m^3/day x 365 days/yr = volume required/yr; therefore, $\frac{23 \times 10^{12} \; m^3}{365 \times 10^6 \; m^3}$ = approximately 63,000 years.

42. Volkov, "The Call of Lake Baikal," p. 8; Galazii, Baikal i problema chistoi vody v Sibiri, p. 10.

43. Volkov, "A Writer's Notes," p. 14.

44. Ibid., p. 14.

45. Oleg Volkov, "Tuman ne rasseialsia" [The fog has not dissipated], Literaturnaia gazeta, April 13, 1965, p. 2.

46. Micklin, "The Baykal Controversy," p. 490.

47. F. Ia. Shipunov, "Budit le zhit' gory?" [Will the mountains continue to exist?], Priroda, no. 9 (September 1969), pp. 82-86.

48. Volkov, "A Writer's Notes," pp. 14-15.

49. G. Gorin, "Literaturnaya Gazeta's Discussion Club: Before It Is Too Late—Open Letter to Oleg Volkov, Author of the Article 'Lessons of Lake Baikal'," Current Digest of the Soviet Press 19 (January 10, 1968): 12. Translated from Literaturnaia gazeta, December 20, 1967.

50. Charlotte Saikowski, "Soviets 'Battle' for Lake: Pollution in Siberia?," Christian Science Monitor, October 8, 1970, p. 12.

51. Shinkarev, "Tiazhba na Baikal," p. 4.

52. G. Orlov, "Priroda i my: snova o Baikal" [Nature and We: again about Baikal], Literaturnaia gazeta, April 10, 1965, p. 2.

53. Merkulov, "Alarm from Baikal," p. 25.

54. N. Chistiakov and E. Kuznetsov, "Neobkhodimye utochneniia" [Indispensable specifications], Literaturnaia gazeta, April 10, 1965, p. 2.

55. Volkov, "The Call of Lake Baikal," p. 8.

56. Gorin, "Literaturnaya Gazeta's Discussion Club," p. 12; L. Shinkarev, "Generous Gift of Baikal—Plan for U.S.S.R. National Park in Siberia," Current Digest of the Soviet Press 17 (October 20, 1965): 33. Translated from Izvestiia, September 25, 1965, p. 6.

57. Dobronravov, Baikal, introduction; Trofimuk and Gerasimov, "Sokhranit' chistotu vod ozera Baikal," p. 58.

58. Chistiakov and Kuznetsov, "Neobkhodimye utochneniia," p. 2.

59. Volkov, "Tuman ne rasseialsia," p. 2.

60. Volkov, "A Writer's Notes," p. 15.

61. Trofimuk, "Tsena vedomstvennogo upriamstva," p. 2; Trofimuk and Gerasimov, "Sokhranit' chistotu vod ozera Baikal," p. 53.

62. Merkulov, "Alarm from Baikal," p. 25; Trofimuk, "Tsena vedomstvennogo upriamstva," p. 2.

63. Goldman, The Spoils of Progress, pp. 193-94.

64. Volkov, "The Call of Lake Baikal," p. 8.

65. Trofimuk, "Tsena vedomstvennogo upriamstva," p. 2; Trofimuk and Gerasimov, "Sokhranit' chistotu vod ozera Baikal," p. 53; Shinkarev, "Tiazhba na Baikal," p. 4.

66. Trofimuk, "Tsena vedomstvennogo upriamstva," p. 2; Volkov, "Tuman ne rasseialsia," p. 2.

67. For a seismic map of the Baikal region, see Baranov, Atlas SSSR, p. 69.

68. Oleg Volkov, "Lessons of Lake Baikal," Current Digest of the Soviet Press 18 (December 20, 1967): 7. Translated from Literaturnaia gazeta, October 11, 1967, p. 12; Chistiakov and Kuznetsov, "Neobkhodimye utochneniia," p. 2; Galazii, Baikal i problema chistoi vody v Sibiri, p. 21.

69. Trofimuk, "Tsena vedomstvennogo upriamstva," p. 2.

70. Volkov, "Tuman ne rasseialsia," p. 2.

71. Trofimuk and Gerasimov, "Sokhranit' chistotu vod ozera Baikal," p. 54.

72. Merkulov, "Alarm from Baikal," p. 25; Trofimuk, "Tsena vedomstvennogo upriamstva," p. 2.

73. Volkov, "A Writer's Notes," p. 14.

74. Dobronravov, Baikal, introduction.

75. Konstantinov et al., "Baikal Waits," pp. 6-7.

76. Shinkarev, "Generous Gift of Baikal," p. 32.

77. Ibid., p. 32; "First National Park," Soviet Life, August 1966, p. 8.

78. B. Semenov, "People's Control Sheet No. 124: Unhurried Designers," Current Digest of the Soviet Press 20 (December 25, 1968): 24. Translated from Pravda, December 9, 1968, p. 3.

79. Galazii, Baikal i problema chistoi vody v Sibiri, p. 23.

80. Ibid., pp. 23-24.

81. V. A. Kirillin, "On Measures for the Further Improvement of Conservation and the Rational Utilization of Natural Resources," Current Digest of the Soviet Press 24 (October 18, 1972): 4. Translated from Pravda and Izvestiia, September 20, 1972, pp. 2-3.

82. M. S. Protas'ev, ed., Vodnye resursy i vodnyy balans territorii Sovetskogo Soiuza [Water resources and water balance of the territory of the Soviet Union] (Leningrad: Gidrometeorologicheskoe izdatel'stvo, 1967), p. 173.

83. Micklin, "The Baykal Controversy," p. 497.

84. Dobronravov, Baikal, photo number 106.

85. "Literaturnaya Gazeta's Discussion Club: The Biosphere, The Strategy of Protection," Current Digest of the Soviet Press 24 (August 2, 1972): 6-7. Translated from Literaturnaia gazeta, May 24, 1972, p. 10.

86. Galazii, Baikal i problema chistoi vody v Sibiri, p. 24.

87. Semenov, "People's Control Sheet No. 124," p. 24.

88. "V Tsentral'nom Komitete KPSS i Sovete Ministrov SSSR: o dopolnitel'nykh merakh po obespecheniiu rational'nogo ispol'zovaniia i sokhraneniia prirodnykh bogatstv basseina ozera Baikal" [In the central committee of the CPSU and the council of ministers of the USSR: on additional measures to ensure the rational utilization and conservation of the natural resources of the Lake Baikal Basin], Pravda, September 24, 1971, p. 1.

89. "Official Department: Concern for Baikal," Current Digest of the Soviet Press 21 (February 26, 1969): 36. Translated from Izvestiia, February 8, 1969, p. 2.

90. "V Tsentral'nom Komitete KPSS i Sovete Ministrov, SSSR," p. 1.

91. The data for 1949 are either unexplainably low or else the 1951-55 average is high.

92. Saikowski, "Soviets 'Battle' for Lake," p. 12.

93. L. Gribanov, "Time Will Tell," Current Digest of the Soviet Press 19 (December 20, 1967): 8. Translated from Literaturnaia gazeta, November 15, 1967, p. 4.

94. V. Baraev, "The Baikal Whitefish—Drive Poachers Out of the Glorious Sea," Current Digest of the Soviet Press 12 (November 23, 1960): 24. Translated from Komsomolskaia pravda, September 30, 1960, p. 2.

95. B. Moskalenko, "Once More about the Fate of the Lake Baikal Omul," Current Digest of the Soviet Press 19 (December 20, 1967): 9. Translated from Pravda, November 17, 1967, p. 3.

96. Ibid., p. 9.

97. Baraev, "The Baikal Whitefish," p. 24; Semenov, "People's Control Sheet No. 124," p. 24.

98. V. I. Astrakhantsev and B. I. Pisarskii, "Vodokhoziaist-vennye problemy Pribaikal'ia" [Water management problems of the Baikal region], Gidrotekhnika i melioratsiia 20 (July 1968): 73.

99. Dobronravov, Baikal, introduction.

100. M. Podgorodnikov and V. Travinskiy, "Priroda i my: Baikal segodnia" [Nature and we: Baikal today], Literaturnaia gazeta, April 5, 1972, p. 10.

101. Gribanov, "Time Will Tell," p. 8.

102. Dobronravov, Baikal, introduction.

103. Shipunov, "Budit le zhit' gory?," pp. 86-87.

104. Galazii, Baikal i problema chistoi vody v Sibiri, p. 24.

105. Shabad, "News Notes," p. 145.

106. Sharov, "Baikal in Autumn—Man and Nature," p. 28.

107. Rossolimo, Baikal, pp. 90-102; Astrakhantsev and Pisarskii, "Vodokhoziaistvennye problemy Pribaikal'ia," p. 73.

108. Gorin, "Literaturnaya Gazeta's Discussion Club," p. 12.

109. Volkov, "A Writer's Notes," p. 15.

110. Gorin, "Literaturnaya Gazeta's Discussion Club," p. 12.

111. "Official Department: Concern for Baikal," p. 36.

112. "V Tsentral'nom Komitete KPSS i Sovete Ministrov SSSR," p. 1.

113. Ibid., p. 1.

114. Podgorodnikov and Travinskii, "Priroda i my: Baikal segodnia," p. 10.

115. Kirillin, "On Measures for the Further Improvement," p. 4.

116. Podgorodnikov and Travinskii, "Priroda i my: Baikal segodnia," p. 10.

117. Ibid., p. 10.

118. A. Efremov, "On the News Map: Safe-Conduct for Forests—Nature and People," Current Digest of the Soviet Press 24 (October 25, 1972): 29. Translated from Izvestiia, September 30, 1972, p. 6.

119. Semenov, "People's Control Sheet No. 124," p. 24.

120. Shinkarev, "Generous Gift of Baikal," p. 33.

121. "Official Department: Concern for Baikal," p. 36.

122. "V Tsentral'nom Komitete KPSS i Sovete Ministrov SSSR," p. 1.

123. Podgorodnikov and Travinskii, "Priroda i my: Baikal segodnia," p. 10.

124. Ibid.

125. Ibid.

126. A. V. Georgiev, "Meetings of the USSR Supreme Soviet: On Measures for the Further Improvement of Conservation and the

Rational Utilization of Natural Resources—Co-Report by Deputy A. V.
Georgiev, Chairman of the Council of the Union's Conservation Committee," Current Digest of the Soviet Press 24 (October 18, 1972):
8-9. Translated from Izvestiia, September 21, 1972, pp. 4-5.
    127. A. Merkulov, "City of Baikalsk under Construction," Cur-
rent Digest of the Soviet Press 13 (May 24, 1961): 35. Translated
from Pravda, April 25, 1961, p. 6.
    128. Novikov, "Bogatstva Baykala—strane," p. 3.
    129. Merkulov, "Alarm from Baikal," p. 24.
    130. Ibid., p. 26.
    131. P. Shternov, "They Have Sufficient Powers," Current Digest
of the Soviet Press 19 (December 20, 1967): 8. Translated from
Literaturnaia gazeta, November 15, 1967, p. 11.
    132. Shinkarev, "Generous Gift of Baikal," p. 33.
    133. Merkulov, "Alarm from Baikal," p. 26.
    134. Volkov, "A Writer's Notes," p. 14.
    135. Micklin, "The Baykal Controversy," p. 494.
    136. Volkov, "A Writer's Notes," p. 14.
    137. Volkov, "Lessons of Lake Baikal," p. 6.
    138. Ibid., p. 7.
    139. Ibid.
    140. Ibid., pp. 7-8.
    141. Gorin, "Literaturnaya Gazeta's Discussion Club," p. 12.
    142. Dobronravov, Baikal, introduction.
    143. A. Ishkov, "Follow-Up to a Literaturnaya Gazeta Article:
Too Early for Final Conclusions," Current Digest of the Soviet Press
20 (February 21, 1968): 24. Translated from Literaturnaia gazeta,
January 17, 1968, p. 10.
    144. Shabad, "News Notes," p. 145.
    145. Astrakhantsev, "Zashchitit' vodnye resursy Irkutskoi o
oblasti ot zagriazneniia," p. 102.
    146. Shipunov, "Budit le zhit' gory?," p. 84.
    147. V. Goldman, The Spoils of Progress, p. 201. Original
source: Pravda, February 16, 1969, p. 1.
    148. V. Goncharov and A. Iurkov, "Na ozere Baikale" [At Lake
Baikal], Komsomolskaia pravda, August 11, 1970, p. 2. Translation
by William Mandel.
    149. Galazii, Baikal i problema chistoi vody v Sibiri, p. 9.
    150. Ibid., p. 13.
    151. Goncharov and Iurkov, "Na ozere Baikale," p. 2.
    152. "Official Department: Concern for Baikal," p. 36.
    153. Ibid., p. 2.
    154. "V Tsentral'nom Komitete KPSS i Sovete Ministrov SSSR,"
p. 1.
    155. "Official Department: Concern for Baikal," p. 36.

156. "V Tsentral'nom Komitete KPSS i Sovete Ministrov SSSR," p. 1.

157. Podgorodnikov and Travinskii, "Priroda i my: Baikal segodnia," p. 10.

158. Ibid.

159. Ibid.

160. Goldman, "The Convergence of Environmental Disruption," pp. 37-42, "Our Far-Flung Correspondents," pp. 58-66, and The Spoils of Progress, pp. 177-209.

161. Goldman, The Spoils of Progress, pp. 178 and 209.

162. Marshall Goldman, "The Soviet Environment," Environment 15 (May 1973): 30.

163. Goldman, The Spoils of Progress, p. 198. Also, see Podgorodnikov and Travinskii, "Priroda i my: Baikal segodnia," p. 10.

164. "Saving Lake Baikal," Newsweek, November 22, 1971, pp. 52 and 55.

165. G. Kapralov, "Cinema: In Siberia, By the Lake," Current Digest of the Soviet Press 22 (June 9, 1970): 9 and 11. Translated from Pravda, May 14, 1970, p. 6.

166. "Saving Lake Baikal," p. 52.

167. Farley Mowat, The Siberians (Baltimore: Penquin Books, 1972), pp. 45-48.

168. Ibid., p. 49.

169. "Russian Concern Low on Pollution," Columbus (Ohio) Dispatch, October 6, 1972, p. 2a.

170. V. A. Krotov et al., eds., Vostochnaia Sibir' [Eastern Siberia] (Moscow: Akademii Nauk SSSR, 1963), p. 703.

171. Novikov, "Bogatstva Baikala—strane," p. 3.

172. "Krai zapovednyi" [Land preserve], Pravda October 23, 1972, p. 2.

173. "Posty dozornykh na Baikale" [Patrol posts at Baikal], Pravda, October 5, 1972, p. 3.

174. "V Tsentral'nom Komitete KPSS i Sovete Ministrov SSSR," p. 1.

175. "Podpisanie Sovetsko-Amerikanskikh soglashenii" [Signing of Soviet-American agreement], Pravda, May 24, 1972, p. 1, "Sogla-shenie mezhdu Soiuzom Sovetskikh Sotsialisticheskikh Respublik i Soedeinennymi Shtatami Ameriki o sotrudnichestve v oblasti okhrany okruzhaiushchei sredy" [Agreement between the Union of Soviet Socialist Republics and the United States of America on cooperation in the field of environmental protection], Pravda and Izvestiia, May 31, 1972, p. 2. Also translated in Current Digest of the Soviet Press 24 (June 28, 1972): 17.

176. U. S. Department of State, Memorandum of Implementation of the Agreement, p. 3.

7

PLANNING FOR PURPOSEFUL USE
OF THE ENVIRONMENT:
A HUNGARIAN VIEWPOINT
Gyorgy Enyedi

As a result of industrial development and urbanization, the natural environment, which ensures the survival of man, has deteriorated to such an extent that the problem is no longer of a local nature. It no longer is a problem only of the large cities and their surroundings, because the pollution of the oceans and the atmosphere touches every part of the world. In 1968 U Thant made his famous report to the General Assembly of the United Nations in which he called for international unified action for the protection of the human environment. Since then this document has been the basis for many international meetings, as well as coordination of research and legal regulation.

The general public is well acquainted with examples of environmental deterioration, and, as a matter of fact, so much attention has been focused on these examples that some people view the various calls for environmental protection as a fad. However, it is safe to say that the seriousness of the situation demands that more attention be given to the problem. According to one statistic, 129 million tons of dust fall yearly on the population of the United States. In the Federal Republic of Germany 2.5 million tons of dust, 5 million tons of SO_2, and 2 million tons of CO_2 poison the environment.

Similar examples also can be cited from Hungary. Hungarian industry annually increases the amount of pollutants in the water supply. While in 1940, 600,000 cubic meters per day of polluted water was emitted, by 1965 this amount had grown to 1,200,000 cubic meters per day, and by 1970 it had reached 1,540,000 cubic meters

This article is an updated version of an earlier study that appeared in the Hungarian sociological and literary journal, Valosag.

per day. Of the continually growing industrial wastes 70 percent
consists of pollutants from the chemical, rubber, iron, paper, and
food processing industries. The pollutant materials from these industries include those that are most harmful to the living environment.

The growth in the quantity of polluted water was not matched by
increases in the capacities of the cleaning apparatus designed to check
pollution. According to representative samples, only 27 percent of
all waters with industrial wastes reach the streams after a proper
degree of cleaning. As a result, the quality of some of Hungary's
streams is so deteriorated that periodically they are not useful for
anything. The most polluted rivers are the Danube (between Budapest
and Dunaujvaros), as well as the Szamos, the Sajo, and the Zagyva,
all tributaties of the Tisza.

The air pollution in the capital and in many other industrial
cities is very severe. In Budapest there is an antiquated heating
system that utilizes coal containing a great deal of sulfur, and the
average value of sulfur dioxide in the city's atmosphere during 1969
was 0.41 mg/m^3 contrasted with the average international norm of
0.15-0.20 mg/m^3. During the winter months this value reaches 1.00
mg/m^3.

These are only a few of the many examples of environmental
deterioration both in Hungary and elsewhere. Mankind has to be
liberated from the threat of polluted water, poisoned air, and other
man-made disasters as a precondition for the realization of those
great prospects offered by the scientific-technological revolution.
The four horsemen of the apocalypse accompanying society along its
path of development are war, famine, conquest, and the destruction
of the natural environment. Only if we succeed in conquering these
threats can we think meaningfully about the distant future.

All activities of mankind, whether they are destructive or cre-
ative, cause a deterioration in some element of the natural environment, and by these activities man decreases the conditions of his own
existence. Since the relationship between man and nature takes place
by the process of action-reaction, the solution to our problems lies
either in stopping man's harmful actions or in finding means to reverse
the harm already done. Although this formulation is too simple, all
of the basic problems related to environmental deterioration must
be dealt with in this way.

Among specialists it frequently has been stated that Hungary's
late industrialization and urbanization gave us a better chance to
deal with the relationship between society and nature. In addition,
it has been presumed that the natural environment of Hungary is less
harmed than those of West European countries that industrialized
centuries before Hungary did. Further, in Hungary the concentrated
industrial-urban damage is occurring in a period when the technical

means to prevent deterioration are well developed and when socialist planning methods give an opportunity for the organization of our entire society against pollution. It should be mentioned, however, that the potential advantage of socialist planning can only be realized if we can fit into the planning system the relationship between man and nature. This incorporation is fundamentally an economic problem. It should be emphasized that the solution to the conflict between man and nature cannot be reduced to solutions of the problem of biological deterioration and to technical defense. In order to find the correct solution we have to find the best use for the natural environment and to develop planning for the economical use of the existing environment.

THE CONCEPT OF NATURAL ENVIRONMENT

The notion of "man and environment" is very elusive and inexact, since the relationship between the individual and his environment should be studied from a different perspective than the study of society and nature. Man's environment does not include nature alone, but also his relationship to his society. The destructive environment acting upon the individual during the industrial development also acts upon the individual through his societal environment. The methods for defense against or improvement of a destructive social environment and a destructive natural environment obviously are different. The only similarity between noise and water pollution is that both are harmful and worsen the quality of man's life, but if we want to avoid these, we have to use entirely different organizations and decision-making mechanisms. For that reason, we frequently have to separate the two different kinds of environments.

This discussion is only concerned with the relationship between society and the natural-geographical environment. According to the geographical definition, the "geographical environment is that part of nature which is known and used by society." Consequently, the natural environment changes together with society, widens vertically and horizontally, and it is not so static or slowly changing as is frequently attributed to the geographical changes that occur slowly during the course of centuries. Therefore, the natural environment, the geographical locale, refers to "societal nature," and thus it is both a natural and historical category. Consequently, we cannot talk of a locale that has retained its original characteristics in any of the developed states, among which Hungary must be included.

During its entire historical activity mankind has played an increasing role in affecting the natural environment, a role that has generally been of negative character. Through this process man intervenes in the natural processes of nature and tends to destroy nature's equilibrium. Until recent times, aside from a few large-scale destructions of forests, nature has been able to correct the destruction of mankind. During the recent decades, however, mankind's activities have been enlarged in scope and quantity and nature frequently is unable to reestablish its own balance. For example, when rivers are polluted beyond a certain extent, they are unable to clean themselves of the waste products. This intensified human activity partially is due to the growth of human population and partially to the population concentration in urban environments. These urban environments are the collecting points of the activities that are destroying nature.

The damage caused by the overly intensified use of environment is threefold. It is of a biological nature (that is, it harms the condition of human existence and the condition of existence of other living beings), it is of a societal nature (it harms man's societal environment such as his place of residence), and it is of an economic nature (it harms the quality of energy sources, thereby creating material harm or increasing the cost of production).

The increased activity of society does not merely harm nature, it also begins such natural processes that, from time to time, destroy the equilibrium of nature and create new equilibria. Even within the human locale, natural processes exist upon which human activities have no effect. Such activities, however, continually decrease. The most well-known of such processes are labeled natural-anthropogenic. These are natural processes, but eventually they affect human activities, or due to human activities they go through quantitative and qualitative changes. For example, they may be speeded up enormously as in the case of soil erosion. Another group of changes is purely anthropogenic. These are the alterations that occur in the natural environment as a result of society's building activities.

We cannot emphasize enough that every portion of the natural environment takes part in those processes that alter nature in its entirety. The human intervention affects the earth cover, the soil, the water, the atmosphere, the flora, and the fauna alike. When mankind intervenes in any portion of these processes—as, for example, altering the chemical composition of the water—it reverberates and affects all other elements of nature and alters the temporary balance of nature. Consequently, from the viewpoint of the control of natural processes, it is never enough to control one single element or to defend against a single agent of harm. It is possible that the alteration in one element is of little or no importance, but as a result of the unstable nature of the environmental balance, even this alteration will affect the total equilibrium. Frequently, for example, in an attempt to defend against environmental deterioration of the soil, man will cause deterioration in unexpected places such as in the water supply and the flora. Very frequently those initiating the technical intervention do not recognize this problem, and only immediate, local problems are considered in trying to correct single causes of deterioration.

THE RELATIONSHIP BETWEEN SOCIETY AND THE NATURAL ENVIRONMENT

The natural environment is an important element in economic growth. It provides food for the population, natural resources for industry, and territory for growing cities. It is clear that along with the growth of human population and with the advent of the scientific-technical revolution, the relationship between society and nature also will continue to broaden and will be further intensified. Society alters new virgin territories in order to get untouched natural resources, and the territories of cities, airports, and industrial locales continue to increase. These are the natural results of economic growth. It would be naive to expect the restoration of the destroyed natural balance from a mere promise that in the future we will curb societal activities. The reason why this point should be emphasized is that already there are opinions that have identified the issue of conservation with the proper utilization of the natural environment, but the former is merely a small part of the latter problem.

Even in the future, natural processes will be altered by societal intervention. Therefore, we can only postulate our purpose to be discovering the direction of the changes and, knowing the predetermined direction of change, to take such measures that speedily and effectively will reestablish the correct natural balance.

THE PLANNED USE OF THE ENVIRONMENT

Environmental protection is a defensive action. Deterioration has already occurred and we are trying to restore the natural balance. The proper use of the environment is an active, planned process: We attempt to discover those processes in the natural environment that take place as a result of the alterations of the societal sphere. We have to attempt to insure a balance between natural and societal processes employing the means of economic planning. This can be handled through an equilibrium analysis within the societal and economic spheres (for example, investments available and investment demands, market supply and demand).

In my opinion, we have to introduce a new type of plan for the environmental use as an integral part of our entire economic planning system. This new type of planning would then serve as a guideline for the future technical intervention affecting our natural environment. To date, the economy of our use of the natural environment is not yet sufficiently known. The intelligent use of the environment demands significant investments and their profitable nature is very difficult to prove. The reason for this difficulty is that we do not have exact, organized information concerning the investments. A second cause is that certain deteriorating activities (for example, the aesthetic deterioration of the environment) can hardly be explained in material and quantifiable terms. A further difficulty is that the negative effect frequently can show up in an economic sector different from the one in which the intervention occurred. For example, the consumers generally are not directly involved in the prophylactic investments that are required for intelligent use of the environment. At the present time we are beginning to enact various laws and regulations forcing those economic and other organizations that use the environment to invest in prophylactic measures even though these organizations themselves are not directly involved in the harmful effects caused by the environmental deterioration. The power to enforce the regulations, however, is minimal; the fines assessed for the violation of laws do not even account for a small portion of the cost of preventive investments needed to combat the deterioration of nature. Therefore, it will be necessary to change and develop the entire regulating mechanism for the intelligent use of the environment. In the present system the fines and administrative decrees can play a role, but they cannot regulate the entire system.

There are many arguments concerning the question whether the investments resulting from the intelligent use of the environment (water purification, defense against soil erosion, etc.) can ever be recovered or whether they should be regarded by society as a necessary sacrifice; however, it is clear that only through such investments can normal conditions of life for future generations be assured. Naturally, single investment decisions still have to be made on the basis of a concrete investment-profit analysis. The profitability of investment, therefore, can influence the priorities of investment as well. Perhaps the question of profitability will appear more relevant if we look at the result of that investment through the entire natural processes and examine its effect on the total environment. If such an effect is quantifiable at all, it is probable that the returns will be similar to those investments made in infrastructure rather than to agricultural or industrial investments. Thus, an investment can be profitable on the level of economic well-being of the nation, although the investor rarely will find his investment returning a profit to the unit. Consequently, we have to make certain that these investments are supported by the government to an increased extent so that the profitability of individual enterprises is not affected.

Planning for the intelligent use of the environment is closely related to planning of the territorial unit. At the same time, individual branches in their planning naturally also are involved with some element of environmental use. In individual branches the decisions, however, appear in some concrete form (for example, in a new industrial plant or in the form of an irrigation system), and there they have an affect on the natural environment. It is imperative to plan ahead for the joint result of these individual efforts. The territorial plan contains the complex economic program of a given area—of a district or an urban agglomeration—and thus it is capable of measuring the total impact of all changes on the environment. It is for this reason that we think it necessary for these territorial development plans to have an environmental component as well. Up to the present time such a plan, or portion of a plan, has been unknown. Even the excellent regional plan designed for Lake Balaton, in which the quality of the natural environment is a primary economic factor, did not include such planning.

Naturally, we are aware of the fact that it is much easier to mention all this in a program than to realize it. It is very difficult to bring together those natural scientists and planning specialists who are capable of recognizing natural processes. The planned use of the environment is a problem that can only be solved by a wide-ranged interdisciplinary cooperation and by a research organization widely divergent from the traditional. We must achieve this goal, however, because the potential advantage of socialist planning can only be realized through such cooperation.

In the natural environment of Hungary, many of the problems of environmental deterioration originate outside the political borders of the Hungarian state. It is well known that 96 percent of our surface water originates outside our borders, and frequently it arrives in seriously polluted condition. The planning for the intelligent use of our water can only be effective if it takes place through international coordination. The new type of cooperation among the socialist states also has to be expanded to include planning for environmental use. Although there has been international cooperation in several industrial branches, we do not at present have an agreement even with our friendly neighboring socialist states in reference to the cleaning of waters and to the division of water supply. In the development of the border zones of the individual states, we are beginning to see international cooperation and we need the same cooperation in the planning for environmental use.

The solution to the problems of society and of nature cannot be constrained to mechanical or biological activities; we also have to have economic solutions. Although they are less precise than the mechanical solutions, they certainly are equally important and cannot be neglected.

CHAPTER

8

**PERCEPTION OF THE
AIR POLLUTION HAZARD
IN LJUBLJANA, YUGOSLAVIA**
David E. Kromm

Pollution of land, air, and water is a world problem that reaches its most serious proportions in the densely settled urban-industrial centers of the more developed countries. Scholars in these nations are devoting an increasingly greater effort toward defining and analyzing the existing levels and causes of deterioration, with the goal of reducing ecological disruption and the resulting negative impact on human well-being. But little is known about the human response to pollution at individual sites or in differing ideological and cultural milieu.

A potentially very useful approach to the social control of environmental problems involves attempts to establish the perception of pollution by populations experiencing high levels of contamination. If we are able to ascertain the awareness of an environmental hazard, the means of coping with its existence, and the feelings about what ought to be done by the community, we can prepare a more satisfactory system of environmental management. Through inventorying the knowledge and attitudes of a population, decision makers are better able to select environmental strategies that are acceptable to a society, more closely reflecting its needs, abilities, and desires. As effective response to pollution is based on human appraisal of the problem, related scientific and technical competence can be best implemented when it is consistent with existing social institutions and priorities.

This chapter considers the perception of environmental pollution in Ljubljana, Yugoslavia, a socialistic, growth-oriented city in Slavic Central Europe. It specifically reports on the attitude of residents toward a serious and worsening air pollution hazard. The determined perception of a random sample of the population suggests several considerations regarding the initiation of measures to cope with the air pollution problem. And in a more general application, the observed behavioral conditions may also have implications for the environmental ethic of all societies. There may well exist a "pollution syndrome" or unique response to a collectivity of pollution hazards that can be identified through examining cross-cultural similarities and differences in perception and adjustment.[1]

PROCEDURES

The procedures employed in this research effort were based on techniques developed by Gilbert F. White, Robert W. Kates, and Ian Burton in their individual and collaborative work on natural hazards. Natural hazards are defined as "the interaction of physical systems and social systems in which there is risk and uncertainty as to extreme events."[2] These men and others continue to investigate variations in perception of natural hazards and the range of individual and community adjustments to them. Although the original methodology was built on human response to flood and drought, it subsequently has been refined and extended to other natural hazards such as urban snow, frost, and tornadoes, and more recently to the quasi-natural hazard of air pollution, which is partially created by man but transmitted through natural processes.[3]

To assess human response and initiative with respect to natural hazards, comparative field observations are being undertaken by co-operating researchers throughout the world. Using a standard set of field procedures and instruments, observations are being made in a variety of cultural and physical conditions. These comparative field observations hopefully will provide new insights into the ways in which man deals with hazards in the environment as well as specific suggestions for improving public policy in guiding such adjustments.[4]

A perception-adjustment questionnaire that is applicable to locally significant hazard conditions has been developed for the observations. The basic interview is modified from place to place to take into account local differences, but key questions are asked at every observation site. Over half of the questions utilized in the Ljubljana inquiry have been or will be used in other areas of high pollution hazard, providing an opportunity for cross-cultural comparison of observational sites. The Ljubljana questionnaire consisted of 57 questions, which identified the interviewee with respect to age, sex, social and economic status, and certain distinguishing personality traits, and elicited responses on his perception of and adjustment to the environmental hazard.

A random sample of 168 was selected in two clusters representing two urban communes of Ljubljana. The sample was obtained through a random draw of household identification cards from commune

131

family files maintained by the city.[5] The sample was drawn proportional to the urban population of the two communes. The inquiry was conducted in June and July 1971 by an advanced student in geography at Ljubljana University. Only one male or female adult in each household was questioned. Through repeated calls, 160 or 95.2 percent of the original sample were eventually interviewed.

SITE DESCRIPTION

With an urban-area population of 270,000, Ljubljana is the largest city in the Republic of Slovenia and is the provincial capital. It functions also as the cultural center for the Slovene people. Ljubljana is an important industrial center with large chemical, machine-building, metal-fabrication, and food-processing plants; and it is a leading transportation node, possessing good rail and highway linkages to Vienna, Zagreb, and the Adriatic coast.[6]

The sources of pollution are closely related to these urban functions. Motor vehicle exhausts are high throughout the year, largely because major highways connecting southern Yugoslavia and the coast with Central Europe pass through the center of the city on three arterial routes. Individual stoves burning soft coal and wood are the leading suppliers of home heating for the rapidly growing population, which largely resides in closely spaced multiple-family dwelling units. This gives rise to highly concentrated maximum cold season emissions for much of Ljubljana. The last major source of pollution is factory emissions, which are greatest in the Moste industrial district.

It is probable that total emissions into the air in Ljubljana are no more than in most Central European cities of similar size. But the poor natural cleansing of the Ljubljana Basin seriously aggravates the air pollution problem. The city is surrounded by mountains or high hills that prevent effective dispersion of the pollution by wind. The situation is worsened in the winter, when temperature inversions occur on most days. During the night there is a rapid loss in heat, resulting in colder air near the surface during the daylight hours. As this cold air is overlaid by warmer air, the home heating, motor vehicle, and factory emissions are kept in the ground layer of air. As fogs occur an average of 150 days each year, the pollution can be especially dangerous.[7]

During the cold season, monthly average concentrations of SO_2, which double the maximum legal concentration of .15 mg/m^3, occur at 14 of the city's 16 measurement stations. At 6 stations, monthly averages at least triple the permitted 24-hour concentration.[8] On January 13, 1967, a concentration of 2.4 mg/m^3 was recorded, which is considerably above the highest measurement in Cincinnati, New

Orleans, or Los Angeles, but below that reached in New York or Chicago.[9] All but one of the monitoring points has monthly SO_2 averages below the maximum in the summer. The smoke levels exceed the statutory limits at 10 of the points in the winter, with all registering acceptable concentrations in the summer.

Measures of pollution vary throughout the city. The densely populated, heavily traversed Center consistently provides the highest readings, followed by adjacent areas in surrounding communes and at monitoring points along several major arterial routes.[10] This study considers perception in two of Ljubljana's five communes (see Table 8.1). The first, Bezigrad, is largely residential, with substantial commercial development only along its major thoroughfare. Measured concentrations of SO_2 and smoke are relatively high where the commune borders the Center, but lower in the more distant and less intensely populated residential areas. The second, Moste, has moderately

TABLE 8.1

Comparisons between Bezigrad and
Moste Communes, 1971

	Bezigrad	Moste
Urban population	30,766	22,810
Sample size	90	70
Percent that own a car	71	71
Percent over age 45	61	59
Percent with very high incomes[a]	14	4
Percent with very low Incomes[b]	7	11
Percent that attended a university or equivalent	31	6
Percent with no high school or equivalent	36	54
Percent perceiving local air pollution problem	39	61

[a]Over 50,000 dinars annual household income.
[b]Below 12,000 dinars annual household income.

Source: Skupscina mesta Ljubljane, Kresija, Prijavni urad.

high measures throughout. It is a working-class area also lying adjacent to the Center, but it has Ljubljana's most important manufacturing district and a large coal-fired steam heating plant further out. These industrial activities emit the greatest amount of visible smoke and particulate matter in the city.

FINDINGS

The issues investigated were awareness of the air pollution hazard, personal adjustment to its existence, and preferred community responses to the problem. A part of the findings on personal awareness and adjustment are considered in this chapter. Under both categories the totals for varying responses will be discussed, often comparing the replies of those living in Bezigrad and Moste. Comparisons between characteristics or attitudes of the respondents and other factors are made whenever there is significant and revealing variation. Unless otherwise indicated, all questions were open-ended or free response, allowing the respondent to express his views as he wished.

Awareness of Air Pollution

Determination of awareness was partly approached through exploring the perception of the distribution of pollution in the city. An early question asked: "Is air pollution a problem in this part of town?" The alternatives were specified, and 78 selected "yes," 15 "no," 56 "doubtful," and 11, "I don't know." Although only 39 percent of those in Bezigrad felt pollution to be a local problem, 61 percent of those living in Moste took this view. As Bezigrad had most of the people with university educations and high incomes, and Moste had a majority of those who did not complete high school and are earning below-average salaries, income and education usually vary in the same manner as place of residence.

The next question asked the respondent: "In which part of town is the air most polluted?" 108 people replied "Moste," 39 "the Center," and 13 gave other answers. Measurement data suggests that the Center has the highest SO_2 and smoke values, with Moste having a high amount of visible factory-induced smoke and particulate matter. Response to this question varied significantly by place of residence. Exactly 33.3 percent of the residents of Bezigrad accurately identified the Center, as compared with only 12.8 percent of those living in Moste. Moste respondents strongly felt their area to be the most polluted.

The third question asked: "In what part of town is the cleanest air?" This resulted in a wide variety of answers, with little difference between the two communes. Of the urban districts, Siska was chosen by 48 respondents and Bezigrad by 28. Outer or suburban areas were named by 35 and an upper-class residential section known as Rozna Dolina by 18. All of these districts have areas that are relatively free of air pollution. Other sections were mentioned by 18 respondents.

Another aspect of awareness probed was perception of the actual sources of pollution. The first question dealing with pollution on the questionnaire read as follows: "We hear a lot these days about environmental problems. I would like to ask you something about the air. Would you tell me in as much detail as you can what the words 'air pollution' mean to you?" 97 mentioned "exhaust gases," 75 "smoke," 16 "bad air" or "unhealthy air," and 6 "dust." 17 gave other answers. The second part of this question asked: "Who are the pollutors?" 102 mentioned factories, 100 said motor vehicles, and 15 gave other responses.

If we assume that in order to perceive something one must first be aware of it, it can be said that those interviewed in Ljubljana were perceiving the air pollution hazard. The people interviewed were highly aware that air pollution existed and possessed a generally accurate notion of the distribution of the problem in the city. The overrating of Moste is not surprising, as this area has many factories that periodically issue large volumes of often dark-colored smoke. This is the most visible source of pollution in Ljubljana, and the residents of Moste are especially conscious of these emissions.

The failure of those interviewed to indicate home heating as a source of pollution suggests inadequate understanding of the origins of the air pollution problem in Ljubljana. Home heating has been established as the leading cause of contamination in the winter. It is possible that by interviewing in the summer, residential emissions were overlooked because of their current unimportance. It might also be that those questioned simply did not see themselves as a part of the problem. Motor vehicles were accurately appraised as a major generator of pollution, but reference was often made to large trucks and buses, not the family automobile. Only the highly visible factories were overrated as a pollution source.

Personal Adjustments to Air Pollution

In considering personal adjustments, an attempt was made to determine how air pollution affects each household, the existing financial losses, and the kinds of actions that each respondent would take to reduce pollution or its negative effects. Two subsequent questions probed the additional distances that the respondents would be willing to walk or ride a bicycle each day to help reduce air pollution.

The first of these questions asked the respondent: "If you really had serious air pollution here, in what way do you think your house would be affected?" As the free response was poor, each interviewee was shown a list of eleven possible affects. Seven items were selected by 90 percent or more of the respondents. Table 8.2 shows the order of frequency of the items. Health was an apparent great concern as also shown by the choice of response to the question: "What effect has polluted air on human health?" 137 selected serious, 13 moderate, 3 little, and 7 said they did not know.

Another question dealt with the perceived impact of pollution and asked: "Does air pollution affect you financially?" 62 replied "yes," 84 "no," and 14 said that they did not know. The second part of this question asked: "In what way?" 32 stated that air pollution soiled their clothing, 20 replied that it made their homes dirty, and 10 gave a variety of other possibilities. No one mentioned health expenses, which possibly reflects the effectiveness of the system of social medicine in removing medical costs as a financial burden.

TABLE 8.2

Appraisal of Household Affects and Individual
Adjustment When Air Is Seriously Polluted

Affects and Adjustments	Percent of Sample:
Most Commonly Selected Possible Affects	
It would cause trouble with breathing	98.8
It would soil the interior of the house	95.0
It would kill garden plants	95.0
It would dirty dishes	93.8
It would adversely affect work and recreation	92.5
It would irritate the eyes	92.5
It would cause poor visibility or smog	91.3
Most Commonly Selected Possible Adjustments	
Stop burning rubbish outside	96.9
Don't hang out the washing	89.4
Switch to a smokeless fuel	88.8
Close the windows to keep pollution out	87.5

Source: Data calculated from author's questionnaire.

Each respondent was told, "Please decide what you would consider doing when the air is seriously polluted," and then asked to reply yes or no to each item on a list of 16 possible individual adjustments. Table 8.2 gives the four alternatives that were affirmatively selected by at least 80 percent of those interviewed: Three items were negatively chosen by over half of the respondents: "Stop using the car," "Write to your deputy," and "Wear a smog mask."

The question asking the respondent to indicate what he would do when the air is seriously polluted followed a very poorly answered one that more closely related to what he was doing. From the interviews it appeared that the people accepted the general problem without relating it directly to their immediate household. Unless they were adjacent to a major source of pollution, the respondents tended to see their specific home as being relatively unaffected by the problem. Only when shown a list of effects did they indicate the possibilities of health, cleaning, and gardening problems. Despite this minimization of the threat, fully 39 percent of those interviewed stated that they suffered financial losses because of air pollution.

Two of the four most commonly selected alternatives from the list of possible adjustments, not hanging the wash outside and closing windows to keep pollution out, are probably adjustments actually being made. The two others, switching to a smokeless fuel and stopping the burning of rubbish outside, may have less meaning. At present, virtually all rubbish is collected by a municipal enterprise. Although many would no doubt prefer smokeless fuels or some form of central heating, a few are in a position to obtain them. Those living in apartment houses have little control over the type of heating employed and those residing in private homes have little effective choice because of the growing shortage of natural gas and relatively clean fuel oil. In the ten-year period preceding the interviews, 35 percent of the respondents indicated that their households did change heating systems, most from coal to fuel oil or linkage with a central heating system. The choice of switching to a smokeless fuel is not consistent with the low emphasis placed on home heating as a source of pollution, but probably reflects a desire for a cleaner and more convenient method of home heating.

The two questions on the willingness of the respondents to increase their walking or riding of a bicycle produced a variety of responses (see Table 8.3). The first question stated: "How much further would you be willing to walk each day if you knew that this would reduce air pollution?" and the second read: "How much further would you be willing to ride a bicycle if you knew that this would reduce air pollution?" Alternatives of 500 meters, 1 km, 2 km, 3 km, more than 3 km, and no further were listed for each. Altogether 69 percent asserted that they would walk an extra 2 km or more, and

TABLE 8.3

Number of Respondents Indicating Willingness to
Walk or Ride a Bicycle to Further Reduce Pollution

Distance of Kilometers	Walk	Bicycle
0.5	3	0
1	47	1
2	61	9
3	35	37
Over 3	14	95
No further	0	18

Source: Data calculated from author's questionnaire.

83 percent said that they would ride a bicycle an additional 3 km or
more if it would help reduce pollution.

The responses did not vary significantly according to perception
of the existence of a local air pollution problem or by ownership of a
car, but were closely related to the age of the respondent. Those
under 30 and between 30 and 45 were generally willing to walk 2 km
or more and ride 3 km and more. Respondents over 45 asserted that
they would walk an additional 1 or 2 km and ride 2 or 3 km. Many of
the older people indicated that they did not ride a bicycle.

IMPLICATIONS

Other studies have shown that hazard awareness is related to
past hazard experiences and the stimulus of written and verbal information. In the most popularly read newspapers and magazines in
Ljubljana, air pollution is not yet a major topic of discussion. A review
of the morning newspaper, Delo, for the period June 1-September 1,
1971, revealed no articles dealing specifically with the problem.
Other aspects of local, national, and world environmental deterioration
were occasionally discussed. The low level of general opinion stimula-
tion via mass media enhanced the probability of awareness being related to actual contact with the hazard and other forms of direct experience.

Perception of air pollution as revealed by the words and actions
of those interviewed was related primarily to place of residence,
which, in the examples given, also represented income and education.

Personal experience appeared to influence awareness, as shown by the great attention given to highly visible factory smoke. This resulted in most respondents overemphasizing the pollution in Moste and underestimating the home heating and motor vehicle exhaust-induced contamination of the Center. The individual adjustments considered did not vary significantly except for the willingness to walk or ride a bicycle, which was largely a function of age. Apparently the long-shared experience with a chronic hazard has resulted in minimal variations in awareness and personal adjustment.

Government officials in the Republic of Slovenia and the City of Ljubljana have not yet internalized the pollution issue. More jobs, a higher standard of living, and, for some, the nationalities question are far more popular causes. Because of the concern with the economy, administrators outside of the health and recreational planning agencies are not likely to be responsive to the findings of a questionnaire such as this. To many this is common sense information about a problem that is not critical. Discussions with numerous city and republic officials indicate that air pollution will not receive much more attention until it gives rise to a serious health crisis. Few now feel this crisis to be imminent. Nonetheless, this study is suggestive for policy in reducing the negative impact of air pollution.

The implications for policy resulting from the part of the study considered in this chapter relate to programs for developing greater awareness of the air pollution hazard and for reducing emissions in the city. Although those living in areas of higher visible contamination were more sensitive to the existence of the problem, they were no better informed as to its sources and distribution. Through the press, TV, other media, and the schools, a public information effort should be initiated that would portray the major causes and characteristics of air pollution by season and in different areas of the city.

Once aware of the importance of the individual household as a source of contamination, there might be greater popular support for expensive programs to reduce home heating emissions and individual adjustments such as limiting the use of the family car under certain circumstances. The apparent willingness of those interviewed to walk or ride a bicycle further suggests that the residents of Ljubljana would respond positively to an effort to relate successful pollution abatement to the individual. The effectiveness of such an effort depends on the relative saliency of the air pollution issue for both individuals and public officials.

Earlier in the chapter I mentioned the possibility of a pollution syndrome that involved a collective response to pollution hazards irrespective of cultural, political, technological, and economic differences between the world's decision makers. There is most certainly a universality about many aspects of human behavior, which raises the question regarding similarities in how man deals with his environment. As the socialist economic model theoretically includes the costs of pollution in its planning as compared with the external treatment inherent in the market economy, the relevance of social guidance is of special interest. A further question regards similarities and differences in human response to quasi-natural and natural hazards. Comparisons of the findings of various hazards studies should shed light on these propositions.

NOTES

1. This "pollution syndrome" is equivalent to the "natural hazard syndrome" postulated by Gilbert F. White. See Progress Report: Collaborative Research on Natural Hazards, Department of Geography, University of Toronto, Toronto, Canada, August 1969, pp. 5-6. In another cross-cultural study of environmental perception, Sonnenfeld noticed many similarities among Alaskan natives and people living in Delaware and stated that "it could be that certain preferences and attitudes toward environment will prove universal." Joseph Sonnenfeld, "Equivalence and Distortion of the Perceptual Environment," Environment and Behavior 1 (1969): 97.

2. Letter from Gilbert F. White, February 8, 1972.

3. Ian Burton and Robert W. Kates, "Perception of Natural Hazards in Resource Management," Natural Resources Journal 3 (January 1964): 412.

4. Gilbert F. White, "Comparative Field Observations on Natural Hazards," paper presented at a special session on Natural Hazards Research: Theory, Methodology, and Policy at the Annual Meeting of the Association of American Geographers, Boston, April 21, 1971, p. 10.

5. Skupscina mesta Ljubljane, Kresija, Prijavni urad.

6. David E. Kromm, "Perspectives on the Slovenian Republic of Yugoslavia," The Geographical Bulletin 3 (November 1971): 43.

7. Bojan Paradiz, "Nekaj karakteristik onesnazenja zraka v Ljubljani," Razprave, Drustvo Meteorologov Slovenije 12 (June 7, 1970): 57.

8. Calculated from data provided by the Hidrometeoroloski Zavod, Socialisticna republika Slovenija, Ljubljana, September 1971.

9. Komisija za proucitev vprasanj s podrocja varstva ozracja pred onesnazenjem, Informacija o problemih onesnazenja zraka v SR Sloveniji (Ljubljana: Socialisticna republika Slovenija Republiski Sekretariat za Urbanizem, 1970), p. 3.

10. Paradiz, p. 62.

9

ENVIRONMENTAL DISRUPTION
IN EASTERN EUROPE
Leslie Dienes

A highly controversial notion of comparative social studies is
the assumption of a broadly universal highway to modernization. Some
societies may never take the road; others may not move far on it; and
some may execute exhausting, erratic detours on the way. But along
the highway they experience quite similar transformation and are subject to similar tolls at the gates. The application of this "convergence
theory" to the evolving pattern of social order under different political
systems may be strictly limited. It may be, as Allen Kassof suggests,
"that, although there are inherent limits to the kinds of societies . . .
compatible with the [urban]-industrial mode, these limits are very
broad indeed."[1] There is little doubt, however, that the costs and side
effects of modernization, stemming from an explosively developing
technology, have been strikingly similar under widely different political
ideologies and organizations.

Environmental disruption is just such a toll along the road to
modernization in general and urban-industrial growth in particular.
It is no secret and probably no longer a surprise that planned socialist
economies also experience environmental deterioration and disruption.
But the magnitudes of these changes, the mechanisms of disruption
and the societal responses to such deterioration are still little known.
Our knowledge of these problems in the People's Democracies of Eastern Europe is particularly meager and this chapter is intended as an
exploratory effort. My research is restricted to Hungary, Poland,
and Czechoslovakia, and to areas where damage to the environment,
and hence to man, has been most acute.

Smog was first reported in Budapest only in 1958. In the follow-ing five years it occurred 17 times.[2] Today, smog sits over the Hungarian capital 20 to 25 days each year, inflicting an estimated 800 million forints ($27 million) damage, not counting cost to human health.[3] Recent measurements at two stations in the metropolis indicate a 7 to 8 percent reduction in solar radiation compared to control points outside the city limits. From November to March, radiation is reduced by over 15 percent.[4] In a few provincial cities the quality of the atmosphere is even worse: every square kilometer of Tatabanya receives nearly 400 metric tons of dust and ashes, together with significant amount of sulfur dioxide.[5]

Air pollution also is very serious in Poland and Czechoslovakia and has been described as catastrophic in such bastions of heavy industry as Upper Silesia and the North-Bohemian coal field. Solar radiation in Warsaw is reduced annually by 5 percent, but in the Upper Silesian Industrial District (GOP)* by well over twice as much.[6] Average daily deposition of particulate matter on the land surface of the conurbation was claimed to equal that for London and Berlin combined, with almost 3,000 tons being deposited in the course of 24 hours.[7] In over ten major cities in the district, the measured amount in the mid-1960s ranged from just under 500 tons per square kilometer to over 1,200 tons per square kilometer per year. The sulfur dioxide content of the air frequently exceeds the designated safety limit of 0.25 milligram per cubic meter several fold (See Table 9.1). The domestic use of sulfurous coal greatly increases the emission of harmful matter, especially in winter, while changing air currents and precipitation results in pronounced and often erratic variation in deposition.[8]

In 1963 the Czechoslovak State Commission for the Development and Coordination of Science and Technology estimated that in the North Bohemian (Most-Chomutov) brown coal basin environmental damage caused by atmospheric pollutants from power stations alone exceeded 1.5 billion Czech crowns each year. These stations produce about 5.64 billion kwh of electricity annually at a self cost of 0.18 crowns per kwh.† Therefore, environmental costs per kwh (1.5/5.64) well surpass the production cost of power to the plants and the true price of power to the national economy is somewhere between 0.40 to 0.50 crowns per kwh. The present selling price is set much lower than that, and at such rates power stations cannot afford more effective

---

*Abbreviation for Gornoslaski Okreg Przemyslowy.

†Self cost equals the cost of all inputs an enterprise must pay for.

TABLE 9.1

Atmospheric Pollution in the Upper
Silesian Cities, 1964-65

	Particulate Matter Settling on Surface (tons per square kilometer per year)	Sulfur Dioxide in Air (milligram per cubic meter of air)	
		Minimum	Maximum
Chorzow	1,243.00	0.02	1.27
Swietochlowice	972.00	n.d.*	n.d.
Siemianowice	661.80	n.d.	n.d.
Zabrze	592.08	0.001	1.15
Bytom	568.08	0.007	0.98
Ruda Slaska	555.36	0.003	0.93
Sosnowiec	520.20	n.d.	n.d.
Bedzin	518.28	n.d.	n.d.
Myslowice	496.20	0.006	0.95
Katowice	482.40	0.007	1.29
Gliwice	n.d.*	0.01	1.56
Dabrowa Gornicza	n.d.	0.006	1.33

*n.d. = no data.

Source: Stefan Zmuda, "Vliv hospodarske cinnosti cloveka na geograficke Prostredi na prikladu Hornosleszke Prumyslove Oblasti" in Teorie a metody vyzkumu Ostravaska prumyslva oblast ve 20 stoleti. Czech translation of Polish original, presented at the November 1966 Opava symposium. Slezský ustav CSAV, Opava 1967, pp. 125-26.

pollution control.[9] The quantity of pollutants continue to grow very fast. Between 1965 and 1970, the emission of sulfur dioxide and hydrogen sulfide in Czechoslovakia increased as fast as total industrial production and the discharge is expected to be double the 1965 amount by the end of the present decade.[10] Atmospheric impurities settling on the surface by means of gravity or precipitation pollute soil and ground water, compounding the harm already caused. In the Most-Chomutov coal field, the worst affected region of Czechoslovakia, life expectancy has been reduced three to four years below the Czech average,[11] and in Polish Upper Silesia the milk yield of cows has reportedly dropped by more than one half.[12]

WATER

Misuse and mismanagement of water resources are widespread and international cooperation for the control of effluence is as yet nonexistent. Hungarian industry treats a mere half of its contaminated industrial water and cleanses only a sixth to sufficient degree. The situation in Budapest is still more alarming: of the city's 1.2 million cubic meters of sewage water, two-thirds are discharged into the Danube without any purification and much of the rest is only partially treated.[13] The problem is aggrevated by the fact that 95 percent of Hungary's surface flow originates outside the country's boundaries[14] and some rivers arrive in an already polluted state. The small streams in the Northern Industrial Region are the most seriously affected since at the border they already carry the unassimilated wastes of the large East Slovakian Iron Works and various chemical plants. No significant cooperation from Czechoslovakia is in evidence so far. That Yugoslavia is as yet less afflicted by upstream Hungarian wastewater is no doubt due to the unindustrialized nature of Hungary's southern half and the far larger flows of the three rivers that cross the Yugoslav border.

In Upper Silesia the great concentration of mining and industry results in serious interference with both surface and subsurface hydrology. The nature and extent of these changes are conceivably different on the two levels, but paucity of information restricts attention to surface water. The Industrial District is situated on a watershed between the Vistula and Odra rivers and has been chronically short of water. Its streams deliver to the Vistula and Odra some 15 billion cubic feet of water per year,[15] one-tenth of the combined mean flow of the rivers supplying the Ruhr. Even on a per capita basis, less water is available than in the latter region,[16] and annual stream flow is claimed to provide barely three-fifths of the required amount.[17] The rest is brought in via pipeline from the peripheries where a system of reservoirs has been constructed. The very large quantities of mine water and urban-industrial effluence discharged equals 80 percent of mean river flow into the Odra and 110 percent of the very much larger flow into the Vistula system (see Table 9.2).[18] (Generally, to be suitable for reuse, the amount of waste discharge should be diluted by at least eight times as much river water.) The basically radial pattern rules out the approach so successful in the Ruhr, where the smallest of the three principal streams, with roughly parallel courses, has been converted into the cloaca maxima of the area, greatly reducing the burden on the other rivers.[19]

Two-fifths of the waste discharge in Upper Silesia is mine water (pit and washery outflow), which contains large amounts of mineral salts (sodium chloride, sodium oxide), the rest is urban and industrial sewage. The latter carries much nondecomposed organic refuse, soluble compounds, and insoluble inorganic substances in suspension. The effluence of the Industrial District pollutes the Vistula as far as Cracow. It contains ammonia, nitrates, sulfates, chlorides, heavy and light oils, tar, iron compounds, and phenols.[20]

Czechoslovakia's record of waste-water treatment is no better. Annual discharge into the Labe, Odra, Morave, and Danube alone equals half the mean flow of the Vltava at Prague. Less than 20 percent of industrial sewage is treated in any fashion.[21] The situation is most critical in the Most-Chomutov and Ostrava industrial districts, regions with scarce water resources, where—as in Polish Silesia—the scanty surface flow is unable to handle the vast amount of urban-industrial effluence.

LAND, SOIL, AND VEGETATION

In direct relationship with his increased technological power, man molds and sculptures the earth's surface. From a whole series of economic sectors involved in this process, only the most significant, namely mining, is dealt with here. Mining produces both convex and concave landforms in the terrain. On the one hand, huge amounts of waste heaps are accumulated; on the other, the land may crack or subside because of removal of minerals (solid or liquid) or disturbance of the originally existing physical and hydrological balance in the geological layers. Though the active shaping of the land surface in Eastern Europe through mining predates the socialist era, just as the abuse of air and water, the scale has vastly increased and the process has accelerated since World War II.

In Upper Silesia, the surface-molding effect of mining and industry is accentuated by the extreme horizontal and vertical concentration of mineral deposits, primarily coal. The Polish Upper Silesian Basin, which cumulatively has produced almost as much coal as the Donbas and half as much as the Ruhr,[22] concentrates six times more coal per square kilometer than the Donbas and 1.8 times more than the Ruhr.[23] In addition, the lead-zinc ores of Upper Silesia probably represent the most important nonferrous metal deposits of Europe worked today. Such mineral concentration clearly affects the localization of mining, industry, and population and multiplies their environmental impact. In the central "saddle" of the Basin, coal mining alone has reshaped over half of the natural surface to such an extent that even a mental reconstruction of the original terrain is difficult. The degree of anthropogenic changes decreases toward the south, but in the north, east, and west evidence of marked interference with the natural landform can often be found. And besides mining, limestone

TABLE 9.2

Mean Annual Stream Flow and Waste Water
Quantities in Upper Silesian Rivers

River	Annual Mean Flow (cubic meter per second)	Quantity of Waste Water (cubic meter per second)	Origin of Waste Water in Percentages		Ratio of Waste Water to Mean Flow
			Urban Sewage	Industrial Waste Water	
Tributaries of Odra					
Bierawka	0.90	0.24	2	98	0.28
Klodnica	2.50	2.85	18	82	1.10
Bytomka	0.70	1.54	22	78	2.20
Drama	0.27	0.02	0	100	0.08
Sola	0.50	0.03	67	33	0.06
Tributaries of Vistula					
Gostynia	2.40	0.90	2	98	0.38
Przemsza	9.70	12.00	7	93	1.25
Biala Przemsza	4.50	1.60	1	99	0.37
Czarna Przemsza (without the Brynica)	1.30	4.80	5	95	3.85
Brynica (without the Rawa)	1.10	2.36	6	94	2.15
Rawa	1.70	1.70	25	75	1.00

Source: Stefan Zmuda, "Vliv hospodarske cinnosti cloveka na geograficke Prostredi na prikladu Hornosleszke Prumyslove Oblasti" in Teorie a metody vyzkumu Ostravaska prumyslva oblast ve 20 stoleti. Czech translation of Polish original, presented

and sand quarrying have also caused substantial modification in the terrain.[24] The area of waste dumps grows by 67 hectares annually (of which 50 hectares are due to coal mining), and the surface of total industrial wasteland is reported to be 30,000 hectares, one-eighth of the GOP territory.[25] Such devastation is equally common in Czech Silesia, Ostrava's ten largest waste heaps alone cover 66 hectares.[26] Depressions connected with subsidence are also very common throughout Silesia, with depths of 10 meters in some places. The problem of the economic utilization of these depressions is still unsolved.[27]

As is well known, open-cast mining can destroy an even larger part of the landscape. The devastated area in the North-Bohemian brown coal field currently reaches 1,200 hectares and is growing at an alarming rate.[28] Since World War II, surface mining has been forced to even greater depths (with the newest works aiming at 200 meters) and the ratio of overburden to coal has drastically increased. From a 2:1 ratio in the 1930s it has now reached 5:1, the highest in Europe.[29] To date very little of the area has been restored and the overburden accumulates in huge waste heaps. With current methods, over 40,000 hectares will be destroyed before the seams are worked out.[30] The 40 largest waste heaps alone are reported to cover 110 hectares of land in Czechoslovakia as a whole with another 700 hectares expected to be covered by the end of the decade.[31] Many of the waste dumps are burning from self ignition. Rarely with open flames, they smolder and burn sometimes for decades. Where pyrite is abundant in the waste, sulfur accumulation on the surface is common. Intensive burning prevents the addition of new mine waste and necessitates the start of another dump, prematurely increasing the land surface so covered.[32]

Waste heaps exert a very harmful effect on their environment. They may cause deformation in the ground and their chaotic distribution impedes natural runoff. Their soluble particles are washed out by rain and carried into the soil and the ground-water supply, changing the latter's chemical composition. When on fire, waste dumps pollute the air, and through water and the atmosphere act toxically on the vegetation.[33]

Because of lack of mineral resources, mining operations in Hungary have not yet affected the Hungarian landscape appreciably. The surface mining of bauxite has only transformed very small parts of the terrain so far. However, the recently opened lignite fields along the Matra foothills will result in greater environmental disruption: a 5 to 1 overburden-lignite ratio is expected, and open cast operations will necessitate the relocation of several settlements.[34]

Even where the land surface has not been greatly disturbed, the soils of major mining regions are subject to diverse and far-reaching physicochemical changes. Both directly and indirectly via the biosphere, gaseous impurities (sulfur dioxide, hydrogen sulfide, carbon oxides, nitrogen oxides, etc.) and urban-industrial sewage may change soil properties by altering the chemical composition and biotic content. Soil samples taken in the GOP (in the vicinity of Dabrowka Mala) showed three to eight times as much zinc and much higher proportions of lead per kilogram of dry soil substance than samples taken beyond the range of air- and water-borne pollutants.[35] Locally, air and water pollution has led to high salinity. Acid industrial sewage and mine water has raised acidity in the ground-water and, indirectly, in the soil over wide areas. Soil acidity related to atmospheric pollution gradually increases from the west to the east of the Silesian Industrial District, corresponding to the direction of predominant westerly winds. Acidity is most pronounced in a zone immediately on the eastern boundary of the district. Soil pollution resulting from infiltration of various organic and chemical compounds takes place mainly along rivers and canals draining urban-industrial effluence. The greatest effects are felt to the east and the west of the GOP, in agreement with the direction of water flow into the Vistula and Odra.[36]

As in other settled regions, man had thoroughly altered the vegetation cover of Eastern Europe well before the socialist era, but the negative effects of his activities on forest and plant growth are continuing. Only the influence of air pollution is considered here. Sulfur dioxide and products of its reactions are the most injurious pollutants to plants. They initiate a series of biochemical defects, decreasing the intensity of photosynthetic and respiratory processes and changing the water balance of the plant. Late germination and limited growth is the result, while leaves whiten, pine needles redden and fall off. Around the GOP, the forest is exposed to such toxic influences over a roughly 800-square-kilometer area.[37] In the Most-Chomutov region of Czechoslovakia, 55,000 hectares of forests have been badly damaged through air pollution, a 140-fold increase since 1937.[38] In Polish Upper Silesia the chlorophyl content of sunflowers has reportedly decreased by a factor of three and even more in the case of certain legumes.[39]

THE INFLUENCE OF THE ECONOMIC AND SPATIAL STRUCTURES

Whatever may be the precise link of environmental degradation with our Occidental belief structure and anthropocentric view of the world, a view shared by Communist ideologists, the economic causes of such abuse are now well enough understood. The total cost of the production process, which includes such diseconomics as environmental and social dislocation, surpasses the sum of costs entering into the monetary expense calculus. The consequences of harmful spillovers are thus shifted to the community and are not borne by the producers. In market economies, the neglect of such "externalities" in determining exchange value leads to the underpricing of output and to a consumption level and consumption mix that are far out of line with true social costs. Despite the vaunted "comprehensiveness" of socialist planning, it is clear that harmful spillovers, social-environmental diseconomies, are not included in the cost calculation of firms in Eastern Europe either. As the example from the North-Bohemian region of Czechoslovakia shows, electric power costs are computed without the inclusion of environmental damages, and the selling price of power is far too low for generating stations to undertake pollution control measures (see pages 142 and 143).

Control of the economic structure is, of course, the hallmark of central planning. The leadership of these countries aims at not simply maximizing total GNP, but GNP of a predetermined structure. It could be argued, therefore, that even without internalizing social costs (that is, including them in the pricing calculation of industries and firms), a socialist economy could effect an output mix that minimizes environmental disruption and other harmful spillovers. That socialist countries have long deemphasized consumer goods, especially disposable products, fancy packaging, and synthetics (while providing a rather admirable level of health care), is well known. Therefore, as Marshall Goldman pointed out, there is less to discard, and low labor costs encourage the collection of waste, a lively junk business, and clean streets.[40] Similarly, the shortage of private cars has so far spared these countries from the worst environmental effects of the automobile, though onerous commuting and extreme rush-hour congestion on buses and trams also represent real social costs. Despite the deemphasis of consumer goods and disposable products, however, the output mix of East European countries is not one that would minimize environmental disruption. Since the beginning of socialist planning, the East European countries have emphasized industry, particularly heavy industry, and have neglected services and light manufacturing—sectors palpably "easier" on the environment than heavy industry. Moreover, the autarchic policies of the past, and for the COMECON as a whole even of the present, meant a reluctance to depend on outside sources for fuel and primary raw materials and resulted in intensive exploitation of low-quality domestic resources. Even in such a mineral-poor state as Hungary, the mining sector in the mid 1960s employed 10.5 percent of total industrial labor force, and mining, metallurgy, and power generation combined 19 percent—a far higher share than in any West European country.[41] Not including prospecting for oil and gas, fuel and energy industries received nearly one-third of all industrial investment between 1965 and 1970 and in the 1950s well over two-fifths.[42]

Presently, the East European countries are increasing the share of the service sector, consumer durables, oil refining, and chemicals in their GNP at the expense of coal and primary metals, but petroleum and chemicals are already bringing as many environmental problems as the old-line heavy industries. Excepting Bulgaria and Romania, Eastern Europe is also rapidly becoming motorized. In Hungary, for example, the growth in car ownership has outstripped expectations, with 37 percent of all automobiles concentrated in the capital. Within three to four years, there will be one car for every tenth person in Budapest and by the mid-1980s, one for every fourth or fifth even by conservative estimates.[43] In Czechoslovakia motorization is far more advanced and no doubt will continue to surpass the Hungarian level by a wide margin for quite some time. With the common use of two-cycle engines, the automobile already affects the quality of the air in some East European cities and will do so increasingly in the future.

In theory, socialist planning should also prevent the excessive regional concentration of production and wealth, claimed to be the consequence of chaotic unplanned development under a system of private ownership. And a more even distribution of population, cities, and industry should lessen environmental damage. While a conscious, albeit selective, geographic dispersion into thinly populated but resource-rich areas has been unquestionably part of the Soviet experience, such decentralization is much less evident in the small East European states where the period of central planning also has been shorter. In the first dozen years or so of the socialist era, lip service was paid to the industrial transformation of backward areas and small urban centers. However, the desire for rapid industrial growth gave priority to the expansion of manufacturing in large cities and established industrial regions. Between 1950 and 1967, Budapest and its immediate hinterland received over a third of all Hungarian state investment, though much less of industrial capital outlays alone.[44] Throughout the 1950s, the city of Warsaw and the four southwestern voivodships of Poland (out of a total number of 17) were allocated 60 percent of all Polish state investment.[45] I have shown elsewhere that no radical decentralization of Hungarian manufacturing, two-fifths of which is still concentrated in the Budapest agglomeration, is feasible since the industrial complex of the Hungarian metropolis is distinguished by a high degree of internal linkages, plays a crucial role in the supply of capital goods for the nation as a whole, and its contribution to Hungary' balance of trade in industrial products well exceeds that of all the provinces combined. Industrial development in the provinces will inevitably require substantial capacity expansion in the agglomeration itself.[46] Postwar development in Polish heavy industry is, in Ian Hamilton's words, a classic example of "spatial inelasticity," involving "the increasing centrifugal scatter of industrial growth in new plant locations outside but near existing agglomerations."[47]

As in Western countries, suburbanization and industrial expansion in and around metropolitan centers gobble up agricultural land, create chaotic sprawls, and make proper land management very difficult. The extremely slow pace of housing construction and administrative restrictions on metropolitan residence resulted in unskilled migrants from the provinces inundating the surrounding villages where restrictions on settlements did not apply or were very laxly enforced. During the 1960s, for example, annual population growth in Budapest proper barely reached 1 percent, but it averaged 2.8 percent in the surrounding 44 villages, which had to accomodate some 60,000 migrants.[48] Virtually no zoning regulations have been enforced and land development has been chaotic and made still more so by the increasing proliferation of weekend cottages and shacks.

STATE OWNERSHIP OF RESOURCES AND ENVIRONMENTAL DISRUPTION

Soviet writers insist that for protecting the environment against pollution "a socialist society has undisputed advantages over a capitalist one . . . because [in the former] there is no contradiction between the interests of society and the interests of individuals. . . . In a society with public ownership of the means of production, environmental disruption will invariably be accidental."[49] Though this also is the official position in the European People's Democracies,[50] East European scholars tend to be less dogmatic when discussing the dissonance between the technological and natural environments and the contradictions between the interest of the individual and that of society as a whole. Acknowledging the lack of harmony between nature and the technological milieu in present-day Hungary, one writer feels that by the turn of the century a much greater reconciliation between the two can and must be effected. He admits the existence of a strong proproduction bias in investment allocation even today, which results in a continued neglect of infrastructure, a reluctance to divert funds to pollution control, and impotency in the enforcement of regulations. It is also conceded that at the present stage of socialism, state ownership of the means of production often encourages embezzlement and an indifference to public property. All these contradictions, of course, will be happily resolved once the transition to communism is completed, sometime at the end of the century.[51]

As Goldman remarked, state ownership of resources and productive facilities means that the government is generally "unable to stand aside as an impartial referee between industry and the citizen consumer. . . . There is usually an identity of interest between the factory manager and local government official."[52] This is no doubt the chief reason why recent administrative measures to relocate highly odoriferous and polluting enterprises from Budapest have met with only modest success. During the 1965-70 period, 124 shops and factory units were instructed to relocate, but as of November 1970 only 22 complied, with another 64 reportedly "in process" of moving. The factory units could not have been very large since so far the whole program affected only 8,000 jobs and will affect perhaps three times as much by the end of the 1970s.[53] At any rate, the action often transfers the pollution problems into the provinces, since the firms generally equip their relocated units with technologically obsolete equipment.

The general identity of interest between the government and factory managers to meet production quotas are also reflected in the usually low and the laxly enforced fines on polluters. For the USSR, Goldman reports that fines in the 1960s have ranged from 50 to 100 rubles and, in addition, the financial plan of enterprises drawn up at the beginning of the year frequently make provision for such payments.[5] In parts of Eastern Europe, the penalties may be stiffer. In Czechoslovakia, the payment of fines has proved much cheaper for enterprises than the cost of installing antipollution equipment.[55] In Hungary, no emission norms for air pollutants had been in existence prior to 1972, and it is highly unlikely that penalties against offending firms would have been systematically applied. Emission standards are currently being worked out and for the most dangerous eight or nine compounds were drawn up in March 1973. However, the norms vary immensely according to geographic areas. The whole country is to be divided into "highly protected," "protected," and "other" districts, and the emission norm for benzin, for example, varies from a stringent 1.5 mg per cubic meter of air in highly protected districts to a very generous 240 mg in the third category. Similarly, huge differences characterize the norms for sulfur compounds, chlorine, lead, and other dangerous pollutants. Industries are also categorized according to their impact on the atmosphere and must now pay a yearly environmental protection charge. For the four industrial categories, this charge ranges from 0.1 percent (thousandth) of gross revenue to 1.2 percent. The charge will form a national clean air fund, to be used for investment on antipollution devices. While this belated concern is commendable, the fund, at best, will come to an annual $5 million at the official exchange rate and much less at the rate Hungarians must purchase Western currency and equipment.[56]

GROWTH VERSUS THE ENVIRONMENT:
THE EAST EUROPEAN VIEW

The East European nations are unimpressed by arguments concerning the impending environmental limits on economic development. They regard the pessimistic forecasts of those who argue that productive activities need to be curbed in order to prevent environmental disasters as fatuous and squarely reject the conclusion set out in the controversial MIT publication, The Limits to Growth. Like the underdeveloped countries, they are firmly committed to continued industrial expansion and hold that for all types of environmental damage effective solutions already exist or can soon be invented.[57] Nor is that only the official position. To an average East European, pronouncements by Americans on the evils of the automobile appear gross hypocrisy, while pictures of superhighways and double-decked cloverleafs evoke admiration, not dismay. Whatever Hungarians may care for full-fledged communism promised by the turn of the century, they certainly covet and are determined to attain the ca. $4,000 per capita income forecast to go with it.[58]

Those concerned with man's impact on the environment admit to limits on demographic growth (hardly a problem in Eastern Europe today) and concede the existence of ecological constraints to the type of "extensive" economic growth that in the past tended to disregard ecological questions altogether. However, they perceive equally real economic constraints to environmental protection. In the words of a Hungarian sociologist, the violation of these economic constraints for the sake of the environment is not an alternative that can be entertained: it is "national suicide." "In semideveloped countries, the protection of the biosphere can influence economic growth only insofar as it does not hinder these nations in their effort to reduce their disadvantage in living standard and the efficiency of production."[59] East European scholars, however, tend to be optimistic with respect to the relationship between nature and technology. Correctly, they realize that the relationship between economic development and environmental quality is nonlinear and is not a simple inverse one. The extent the two conflict with and complement each other is influenced by a maze of societal factors, values, and aims. In this optimistic view, it should be possible to advance toward an ecological civilization in which, after the attainment of modest comfort, human wants are increasingly directed toward cultural, health, and social services and grace and quality in life style.[60]

These are brave words. Unfortunately, little practical suggestion is provided for the charting of this course. The "internalization" of social costs in the production process is actually less advanced in Eastern Europe than in the West.[61] As in other fields, East European scholars are looking to Western examples for introducing environmental costs and benefits into industrial and technological growth models.[62] However, the reliability of technology assessment itself depends on modern technology, such as high-speed computers, of which these countries are painfully short.

For the near future I am also skeptical concerning the rejection of the ideal of mass consumption and a shift on the people's preference scale toward a simple but ecologically wholesome civilization. The narrow range of consumer goods and sophisticated synthetics in Eastern Europe is viewed not merely as an inconvenience but also as a sign of backwardness, a sore point for these proud nations as infected by the expansionist, acquisitive spirit of Western civilization as the more developed parts of the Occidental realm. A move toward a non-mechanistic, concentric, essentially medieval view of the universe may be in evidence today in the richest countries of the Western world, satiated with gadgets and high on cholesterol. It is here that organic food and Zen appeals to youth and St. Francis' rejection of man's imperious rule over nature is echoed from scholarly pens.[63] It is too early to expect such a move in Eastern Europe. Still, one hopes that even as these nations have telescoped the abuse of environment into a shorter time span, eventually they will resolve the worst aspects of that abuse with less wavering and delay than countries more advanced.

NOTES

1. Allen Kassof, ed., Prospects for Soviet Society (New York: Praeger, 1968), p. 505.

2. Ferenc Probald, "Budapest varosklimaja," Foldrajzi Kozlemenyek 14, no. 4 (1966): 321.

3. Figyelo, no. 15 (April 14, 1971), p. 1; Magyar Nemzet, March 12, 1971, p. 3; and Figyelo, no. 26 (June 28, 1972), p. 21.

4. Ferenc Probald, "Budapest varosklimajanak energiahaztartasi alapjai," Foldrajzi Ertesito 20, no. 1 (1971): 16-17 and 28.

5. Figyelo, no. 26 (June 28, 1972), p. 21.

6. Jerzy Kondracki, "Sesja naukowa PAN w Szczecinie na temat 'Czlowiek i srodowisko'," Czasopismo Geograficzny 42, no. 2 (1971): 194-202. Quoted in Referativny Zhurnal, Geografiia, no. 4 (1972), L28.

7. Janusz Paszynski, "Investigation of Local Climate in the Upper Silesian Industrial District," Geographical Studies (Polish Academy of Sciences), no. 25 (1959), pp. 88-89.

8. Janusz Paszynski, "Der Jahresverlauf der Luftverunreinigungen im Oberschlesischen Industriegebiet," Angewandte Meteorologie

4, no. 6 (October 1962): pp. 161-65, and "L'influence des conditions climatiques sur le developpement des villes," Geographia Polonica, no. 12 (1968), pp. 97-100.

9. Petr Synek, "Negative Wirkungen der Wirtschaftstatigkeit im Nordbohmischen Kohlenrevier," Ceskoslovenska Akademic Ved, Geograficky Ustav, Studia Geographica, no. 7 (1969), p. 127.

10. An over 33 percent growth in the emission of sulfur compounds versus a 38 percent growth for total industrial production. G. Mikulas, "Exhalaty a boj proti nim," Zivotny prostred 5, no. 4 (1971): 193-99, quoted in Referativny Zhurnal, Geografiia, no. 1 (1972), L27 and Statisticka rocenka CSSR, 1971.

11. Synek, op. cit., p. 127, and Mikulas, op. cit.

12. Paszynski, "Investigation of Local Climate," op. cit., p. 91, and Figyelo, no. 15 (April 14, 1971), p. 1.

13. Figyelo, no. 9 (March 1, 1972), p. 1.

14. Lajos Danicska, "Vizgazdalkodas," Gazdasag 4, no. 3 (1970): p. 81.

15. Sylwia Gilewska, "Changes in the Geographical Environment Brought About by Industrialization and Urbanization," Geographia Polonica, no. 3 (1964), p. 207.

16. The mean annual discharge of the Ruhr river is 82 billion cubic feet, though variations from the mean are great. The Lippe and the small Emscher deliver perhaps an equal amount. The Lippe, immediately north of the Inner Ruhr conurbation, is used extensively today to supplement water supply from the Ruhr River. The Inner Ruhr agglomeration contains a little less than 6 million people and all the cities of the Ruhr Planning Authority (Siedlungsverband Ruhr-kohlenbezirk) —about 10-11 million. However, several of these lie along or west of the Rhine and thus are not dependent on the Ruhr-Lippe water supply. The Upper Silesian conurbation (GOP) has a population of nearly 2 million. Allen V. Kneese, "Water Quality Management by Regional Authorities in the Ruhr Area," in Marshall I. Goldman, Controlling Pollution (Englewood Cliffs, N.J.: Prentice-Hall, 1967), pp. 125-26; Peter Hall, The World Cities (New York: McGraw-Hill, 1966), pp. 122-57; and Czeslaw Kotela, "Main Urban Planning Problems in the Silesian-Krakow Industrial Region," in Jack Fisher, ed., City and Regional Planning in Poland (Ithaca, N.Y.: Cornell University Press, 1966), pp. 111-15.

17. Ryszard Szmitke and Tadeusz Zielinski, "Regional Planning in the Upper Silesian Industrial District," in Fisher, op. cit., p. 307.

18. Stefan Zmuda, "Vliv hospodarske cinnosti cloveka na geograficke Prostredi na prikladu Hornosleszke Prumyslove Oblasti" in Teorie a metody vyzkumu Ostravaska prumyslva oblast ve 20 stoleti. Czech translation of Polish original, presented at the November 1966 Opava symposium. Slezsky ustav CSAV, Opava 1967, p. 120.

19. Kneese, op. cit., pp. 116-20.

20. Gilewska, op. cit., pp. 207-08.

21. Y. Silar and Y. Leden, "Problematika komplexni ochrany a tvorby prirodniho prostredi . . .," Vodni hospodarstvi, no. 1 (1970), p. 4; and I. M. Mayergoyz, Chekhoslovatskaya Sotsialisticheskaya Respublika (Moscow: "Mysl'," 1964), p. 49.

22. Bronislaw Kortus, "Donbas and Upper Silesia—A Comparative Analysis of the Industrial Regions," Geographia Polonica, no. 2 (1964), pp. 186-87; George W. Hoffman, A Geography of Europe. 2d ed. (New York: Ronald Press, 1961), p. 434; Norman J. G. Pounds, The Ruhr (Bloomington: Indiana University Press, 1952), Fig. 12, p. 100; and statistical yearbooks of the respective countries.

23. Kortus, op. cit., p. 185.

24. Zmuda, op. cit., pp. 112-15.

25. Gilewska, op. cit., p. 202.

26. V. Krontilik, Haldove pokryvy na wzemi mesta Ostrava. Slezsky studijni ustav (Opava, 1954).

27. Gilewska, op. cit., p. 205; Zmuda, op. cit., pp. 114-16; and Mieczyslaw Klimaszenoski, "Problems of Geomorphological and Hydrologic Map on the Example of the Upper Silesian Industrial District," Geographia Polonica, no. 1 (1961), p. 77.

28. Synek, op. cit., p. 126.

29. Mayergoyz, op. cit., p. 240; and Ferenc Erdosi, "A banyaszat felszinformalo jelentosege," Foldrajzi Kozlemenyek 14, no. 4 (1966): 327.

30. Synek, op. cit., p. 126.

31. Silar and Leden, op. cit., p. 5.

32. Erdosi, op. cit., p. 336.

33. Gilewska, op. cit., p. 203.

34. Erdosi, op. cit., p. 329; and Tivadar Bernat, ed., Magyarorszag gazdasagfoldrajza (Budapest: Tankonyvkiado, 1969), pp. 528-29.

35. Gilewska, op. cit., p. 209.

36. Zmuda, op. cit., pp. 120-24; and Janusz Paszynski, "Transparence de l'atmosphere comme element du climat local des regions industrielles," Przeglad geograficzny 32 (1960), Supplement: 105-07.

37. Zmuda, op. cit., pp. 128-31; and Gilewska, op. cit., p. 209.

38. Synek, op. cit., p. 127.

39. Zmuda, op. cit., p. 127.

40. Marshall I. Goldman, "Environmental Disruption in the Soviet Union," in Thomas R. Detwyler, ed., Man's Impact on Environment (New York: McGraw-Hill, 1971), p. 72.

41. Kozponti Statisztikai Hivatal, Magyar statisztikai zsebkonyv (Budapest, 1971), p. 99.

42. Ferenc Levardi, "Energiapolitikai koncepcio, 1964-1980," Gazdasag 3, no. 4 (1969): 7-8; and Kozponti Statisztikai Hivatal, Iparstatisztikai zsebkonyv (Budapest, 1962), pp. 90-91.

43. Antal Bacso and Ferenc Janoshegyi, "Megjegyzesek Budapest kozuti kozlekedesfejlesztesi tervehez," Varosepites 5, no. 6 (1971): 16.

44. Jozsef Korodi, Valtozasok Magyarorszag gazdasagi terkepen (Budapest: Kossuth Konyvkiado, n.d.), Table 1, p. 11.

45. Andrzej Wrobel and S. M. Zawadzki, "Location Policy and the Regional Efficiency of Investments," in Fisher, op. cit., p. 437.

46. Leslie Dienes, "The Role of the Budapest Agglomeration in Hungarian Industry," Geographical Review, July 1973.

47. F. E. Ian Hamilton, "The Location of Industry in East-Central and Southeast Europe," in George W. Hoffman, ed., Eastern Europe: Essays in Geographical Problems (London: Methuen, 1971), p. 183.

48. Ivan T. Berend, "Fordulopont es ellentmondasok az urbanizacioban," Valosag 15, no. 12 (December 1971): 12.

49. Igor Petryanov, "Public Greed? No! Public Weal!" Soviet Life, no. 12 (December 1971), pp. 42-43.

50. See, for example, Figyelo, No. 37 (September 13, 1972), p.2.

51. Geza Kovacs, A nagy tavlatok es a tervezes (Budapest: Kozgazdasagi es jogi Konyvkiado, 1970), pp. 149-53 and 166-71.

52. Goldman, "Environmental Disruption," p. 68.

53. Magyar Nemzet, November 11, 1970, p. 3, and November 14, 1970, p. 1; Figyelo, no. 40 (October 7, 1970), p. 4. Substantial government funds are also made available to spur relocation.

54. Marshall I. Goldman, The Spoils of Progress: Environmental Pollution in the Soviet Union (Cambridge, Mass.: MIT Press, 1972), p. 35.

55. Silar and Leden, op. cit., p. 4.

56. Figyelo, no. 15 (April 14, 1971), pp. 1-2, and Magyar Nemzet, March 12, 1971, p. 3.

57. See, for example, Figyelo, no. 32 (August 9, 1972) and no. 37 (September 13, 1972), p. 2.

58. Kovacs, op. cit., pp. 95 and 175.

59. Andras Szesztay, "Kornyezetvedelem es gazdasagfejlesztes," Valosag 15, no. 9 (September 1972), p. 26.

60. Ibid., pp. 28-34.

61. Arthur Wright has suggested that only in the case of perfect centralization and "complete information available to, utilized by, and communicated from the central planning agency is there any reason to expect command planning of the Soviet type to internalize costs not internalized under a market mechanism." But since information is not a free good to planners, the latter "will have considerable incentive to economize information, that is, to apply relative priorities to the selection of information and its incorporation into the planning process." Arthur W. Wright, "Environmental Disruption and Economic Systems," The ASTE Bulletin 13, no. 1 (1971): 12.

62. Zbyszko Chojnicki, "An Economic Approach to Some Problems in Using Geographical Environment," Geographia Polonica, no. 20 (1972), pp. 42-47.

63. Lynn T. White, Jr., "The Historical Roots of Our Ecological Crisis," in Detwyler, op. cit., pp. 27-35.

Design of Cellulose and Paper
  Plants, State Institute for the,
  85
Design of Pulp and Paper Industry
  Enterprises in Siberia and the
  Far East, State Institute for the
  (Sibgiprobum), 86
Development and Coordination of
  Science and Technology, Czecho-
  slovak State Commission for the,
  142
Dmitriev, 61
Dneprodzerzhinsk, 47
Dnepropetrovsk, 45
Dnieper River Valley, 45
Dockstader, Robert, 42
Don River, 55, 57, 59, 60
Donbass, 14, 145
Donbass-Dnieper Bend, 43; air pol-
  lutants in, 45-47, 49
Donetsk, 45
Donitov, Iu. E., 107
Dossov, Comrade, 102
Drucker, Peter F., 37
Dudun River, 105
Duma, 23
Dunaujvaros, 124
Dupont de Nemours, E.I., & Co., 2
Dzhida River, 98

East Siberian: Economic Council,
  86; Fishing Industry Administra-
  tion, 102; State Hatchery Trust,
  101
East Slovakian Iron Works, 144
Eastern Europe, environment prob-
  lems of, 141-154
Economic Commission for Europe
  (UN), 27
Ekologiia, 28
Engels, Friedrich, 28
Enisei: Bay, 83; River, 11, 73, 83
Environment, natural, 125-129;
  best use of, 127-129
Environment and Water Management,

Ministry of (East German), 21
Environmental Protection Agency
  (EPA; U.S.), 19, 21, 23, 25, 43
Erie, Lake, 3, 81, 113
European Economic Community
  (EEC), 19

Fedorenko, Academician, 20
Fedorov, Evgenii K., 112
Fergana, 49
Fishing Industry, Ministry of the
  (USSR), 22, 102, 106, 107, 111

Gainutdinov, Mikhail B., 106
Galati, 5
Galazii, Gregorii, 86-87, 88-90,
  109-111, 113
Geographical Society of the USSR,
  92
Geography, USSR Institute of, 97
Georgiev, A. V., 107
Gerasimov, Innokenty, 86, 109
Germany, 23
Germany: East, 21; West, 29, 123
Giprobum Organization, 97
Giul, K. K., 70
Glavlit (see Literary Affairs and
  Publishing, Main Administration
  for)
Gobi Desert, 92
Goldman, Marshall, 9, 27, 37, 81,
  87, 95, 112-113, 149, 151-152
Goloustnaia River, 83, 103
Goluboi Serdtse Sibiri (Blue Heart
  of Siberia), 97
Goncharov, Anatolii V., 106
Gorin, G., 108
Gornoslaski Okreg Przemyslowy
  (GOP) (see Upper Silesian Indus-
  trial District)
Grand Canyon, 81
Great Bear Lake, 113
Great Lakes, 1, 81
Great Slave Lake, 113
Greater Chivyrkui River, 102

IVAN VOLGYES, Associate Professor of Political Science at the University of Nebraska in Lincoln, Nebraska, received his Ph.D. from The American University in Washington, D. C. He has traveled and completed research in the Soviet Union and Eastern Europe. He is the author of The Hungarian Soviet Republic, 1970, and Hungary in Revolution, 1971, is coauthor of Czechoslovakia, Hungary and Poland: Nations at the Crossroads of Change, 1971 (with Mary Volgyes), Politics in Hungary, 1974 (with Peter Toma), and coedited (with Roger Kanet) On the Road to Communism: Fifty Years of Soviet Domestic and Foreign Policy 1972. In addition, Dr. Volgyes has also published many articles in such scholarly journals as Cahiers du Monde Russe et Sovietique, Problems of Communism, East Europe, East European Quarterly, Journal of Political and Military Sociology, World Affairs, and Choice. In addition to his scholarly activities, Dr. Volgyes is also the Chairman of the Education Committee of the American Association for the Advancement of Slavic Studies.

KEITH BUSH is Senior Researcher at Radio Liberty in Munich, Germany. Dr. Bush who received his M.A. from Harvard in Soviet area studies, is widely respected for his extraordinarily broad interests in the USSR and the many excellent volumes published by that organization for which he has been chiefly responsible. He is also the author of many scholarly articles published in the United States and in Western Europe.

VICTOR L. MOTE is Assistant Professor of Geography at the University of Houston. Dr. Mote received his Ph.D. from the University of Washington and his main interests are problems of air pollution and environmental deterioration in the USSR.

IHOR STEBELSKY received his Ph.D. at the University of Washington and is currently Associate Professor of Geography at the University of Windsor, Ontario. Dr. Stebelsky has published widely and his contributions have been included in such journals as Land Economics, International Geography, and in the Proceedings of the Canadian Association of Geographers.

PHILIP P. MICKLIN received his Ph.D. from the University of Washington where his fields of specialization were geography of the

USSR and conservation. He is currently an Assistant Professor of Geography at Western Michigan University in Kalamazoo. His articles are included in such scholarly journals as the Canadian Geographer, Soviet Geography, and Natural Resources Journal.

CRAIG ZUMBRUNNEN, Assistant Professor of Geography at Ohio State University, received his Ph.D. from the University of Washington. His main research interests are in the field of water pollution in the USSR.

GYORGY ENYEDI, Senior Researcher of the Geographical Institute of the Hungarian Academy of Sciences, received his Ph.D. from the Karl Marx University of Economics in Budapest. He has traveled widely both in the West and in Eastern Europe. He was a Ford fellow in 1967-68 and has traveled and lectured all over the United States. In addition to several dozens of articles in scholarly journals, Dr. Enyedi is also the author of the first volume describing American farming published in Eastern Europe: Farms and Men; The American Agriculture (Budapest, 1971). Dr. Enyedi is the head of the Study Group on Rural Regions of the International Geographic Union and the former Deputy Director of the Geographical Institute of the Hungarian Academy of Sciences. For the last two academic years (1972-74) Dr. Enyedi has been Visiting Professor of Geography at Valery University at Montpellier, France.

DAVID E. KROMM is Associate Professor of Geography at Kansas State University. He received his Ph.D. at Michigan State University and has published widely in such journals as The Canadian Geographer, The Journal of Geography, The Journal of Forestry, and in several scholarly journals in Yugoslavia. Dr. Kromm was on a sabbatical leave in Yugoslavia during the academic year 1972-73.

LESLIE DIENES received his Ph.D. from the University of Chicago. He has published two highly regarded volumes: Locational Factors and Locational Development in the Soviet Capital Industry and The Natural Gas Industry of the USSR; he has also contributed scholarly articles to such journals as Soviet Studies, Geographical Review, and Energy Policy, as well as in Tijdeschrift voor Econ. en Soc. Geografie. He spent the academic year 1970-71 in Eastern Europe and the USSR as a recipient of a grant from the American Council of Learned Societies. Dr. Dienes is currently an Associate Professor of Geography at the University of Kansas.

RELATED TITLES
Published by
Praeger Special Studies

AIR QUALITY MANAGEMENT AND LAND USE
PLANNING
George Hagevik, David Mandelker,
and Richard K. Brail

ENVIRONMENTAL POLICY: Concepts and
International Implications
edited by Albert E. Utton
and Daniel H. Henning

INTERNATIONAL ENVIRONMENTAL LAW
edited by Ludwik A. Teclaff
and Albert E. Utton

THE INTERNATIONAL POLITICS OF MARINE
POLLUTION CONTROL
Robert A. Shinn

MANAGING SOLID WASTES: Economics,
Technology, and Institutions
Haynes C. Goddard